SpringerBriefs in Food, Health, and Nutrition

Springer Briefs in Food, Health, and Nutrition present concise summaries of cutting edge research and practical applications across a wide range of topics related to the field of food science, including its impact and relationship to health and nutrition. Subjects include: Food Chemistry, including analytical methods; ingredient functionality; physic-chemical aspects; thermodynamics Food Microbiology, including food safety; fermentation; foodborne pathogens; detection methods Food Process Engineering, including unit operations; mass transfer; heating, chilling and freezing; thermal and non-thermal processing, new technologies Food Physics, including material science; rheology, chewing/mastication Food Policy And applications to: Sensory Science Packaging Food Qualtiy Product Development We are especially interested in how these areas impact or are related to health and nutrition. Featuring compact volumes of 50 to 125 pages, the series covers a range of content from professional to academic. Typical topics might include:

- A timely report of state-of-the art analytical techniques
- A bridge between new research results, as published in journal articles, and a contextual literature review
- A snapshot of a hot or emerging topic
- An in-depth case study
- A presentation of core concepts that students must understand in order to make independent contributions

Sarfaraz Ahmed Mahesar
Syed Tufail Hussain Shah
Mohamed Fawzy Ramadan
Waris Nawab Panhwar

Authentication of Halal Products using FTIR Spectroscopy

Sarfaraz Ahmed Mahesar
National Centre of Excellence in Analytical Chemistry
University of Sindh
Jamshoro, Pakistan

Mohamed Fawzy Ramadan
Department of Clinical Nutrition
Umm Al-Qura University
Makkah, Saudi Arabia

Syed Tufail Hussain Shah
National Centre of Excellence in Analytical Chemistry
University of Sindh
Jamshoro, Pakistan

Waris Nawab Panhwar
National Centre of Excellence in Analytical Chemistry
University of Sindh
Jamshoro, Pakistan

ISSN 2197-571X ISSN 2197-5728 (electronic)
SpringerBriefs in Food, Health, and Nutrition
ISBN 978-3-031-92772-0 ISBN 978-3-031-92773-7 (eBook)
https://doi.org/10.1007/978-3-031-92773-7

© The Editor(s) (if applicable) and The Author(s), under exclusive license to Springer Nature Switzerland AG 2025

This work is subject to copyright. All rights are solely and exclusively licensed by the Publisher, whether the whole or part of the material is concerned, specifically the rights of translation, reprinting, reuse of illustrations, recitation, broadcasting, reproduction on microfilms or in any other physical way, and transmission or information storage and retrieval, electronic adaptation, computer software, or by similar or dissimilar methodology now known or hereafter developed.
The use of general descriptive names, registered names, trademarks, service marks, etc. in this publication does not imply, even in the absence of a specific statement, that such names are exempt from the relevant protective laws and regulations and therefore free for general use.
The publisher, the authors and the editors are safe to assume that the advice and information in this book are believed to be true and accurate at the date of publication. Neither the publisher nor the authors or the editors give a warranty, expressed or implied, with respect to the material contained herein or for any errors or omissions that may have been made. The publisher remains neutral with regard to jurisdictional claims in published maps and institutional affiliations.

This Springer imprint is published by the registered company Springer Nature Switzerland AG
The registered company address is: Gewerbestrasse 11, 6330 Cham, Switzerland

If disposing of this product, please recycle the paper.

Description

Halal authentication in food is an expanding multifaceted tactic as Muslim consumers' passion for Halal food grows. The development and revenue from Halal food are expected to increase even more in the foreseeable future, generating opportunities for both food producers and traders. Today, the demand for Halal products in the market is significantly growing. Even in non-Muslim countries, we can find supermarkets selling food items labeled as Halal. Hence, Halal authentication is crucial when considering the needs of Muslim consumers worldwide. The mention of Halal as a food authentication parameter pleases consumers wanting to meet religious obligations. Halal food authentication is very important for Muslims in order to obey the Sharia, which was stipulated. Nonetheless, dishonesty has risen, which raises the need for preventive measures. This occurs due to problems with contamination, chemical addition, and the substitution of non-Halal or haram ingredients during the stages of processing or labeling. These difficulties could lead to confusion among Muslim consumers. Therefore, it is crucial to authenticate genuine Halal items, including the establishment of reliable systems to protect the market from haram or non-Halal items.

Reports revealed that Fourier-transform infrared spectroscopy (FTIR) can provide a highly specific and sensitive method to authenticate the presence of bovine serum albumin (BSA) in chili sauce. BSA is not a discernible portion of the formulation of chili items; thus, the existence of BSA endorses this sample as being nonauthentic. Furthermore, a rarely utilized FTIR area indicates typical spectroscopy of unknown dissimilar safeguard mutton with goat fat supplementation in beef jerky products. Findings related to oil adulteration (namely palm oil, palm kernel oil, coconut oil, fish oil, and sunflower oil) were reported. Different proportions of lard have also been studied by using FTIR to a strong consequence. The research base is continually enlarged through the use of novel equipment and tactics that can be efficiently utilized to authenticate Halal food in much more detail. FTIR, an easy method that is nondestructive and non-retrospective, arises as a promising method that is able to authenticate Halal food products. Therefore, a large database covering all similar features taken from various products should be continuously updated.

Importance of Halal Authentication in the Food Industry

Nowadays, the term "Halal" is not only important to Muslims but also captures the attention of the flocks of business food chain operators, scientists, and consumers in general. Islam is the second most widespread religion in the world, and one in every four people is a member of the Islamic faith. The international trade in Halal products, including food and nonfood, has increased significantly in recent years. The Muslim population is forecasted to grow and contribute to consumers' demand for Halal products. However, Halal authentication in the food industry is still considered an issue due to a lack of consensus and a harmonized standard. The non-Muslim population is also interested in buying Halal products, and interest in Halal food extends to non-Muslims for endorsements of safe, hygienic, and healthy food.

Under these circumstances, it can also be said that in order to give confidence to Muslim and non-Muslim consumers and tilt market demand, it is important to have a strong Halal authenticity chain in place. A strict Halal authenticity chain must be implemented, starting from the receipt of raw ingredients to the final products and, finally, the packaging and trading of the final distributed products. A robust Halal authentication process is also required should the authenticity of the product be needed. Currently, Halal authentication is carried out using analytical methods, employing chemical, physical, or biological assays of the food's nature. High-performance, complex analytical methods provide optimal results; however, they are costly, time-consuming, and difficult to operate for Halal authenticity assessment.

Concepts

With the high recognition of religious and dietary purposes of Halal food by Muslims, providing food authentication and certification processes has become inevitable and significant. The need to evaluate and verify the Halal certification to give confidence to the Halal consumer is growing. The failure of the supply chain to verify Halal food authenticity could potentially impact the relationship between consumers and retailers. The use of chemometric techniques in combination with Fourier-transform infrared spectroscopy (FTIR) allows us to obtain rapid, low-cost, and noninvasive analyses, which are still important advances in food authentication for Halal food control. To ensure the authenticity of Halal food, various authentication methods are involved to avoid cheating and mislabeling. The chemometric approach mainly includes principal component analysis, multivariate regression analysis, and partial least squares, allowing us to rapidly determine protein, fat, and carbohydrate concentrations, as well as to differentiate bovine and porcine gelatins present in Halal products using FTIR spectroscopy. In general, chemometric techniques yield rapid and accurate procedures for the development and evaluation of Halal food analytical approaches. These chemometric-based methods are discussed in this book.

Authentication of Halal Products Using FTIR Spectroscopy covers topics with a focus on the latest techniques used for Halal authentication. This is an essential book for food scientists and authorities working in the field of Halal food authentication.

Preface

Many people who are interested in Halal cuisine and the methods for Halal authenticity are the target audience for this book. The main goal of the book is to widen the application of Fourier-transform infrared spectroscopy (FTIR) in conjunction with chemometrics for Halal authentication. The authenticity of food has always been a major issue for producers, regulators, and consumers. There has been a concern about the analysis procedure and the ability of the industry to undertake quantitative measurements that provide statistically sound and unbiased data to underpin risk assessments. Hundreds of discretely identified scientific, regulatory, social, and environmental challenges were reported and reviewed with leading academicians, regulators, and policymakers. Within this group, a number is relevant to food authentication to more specific targets for Halal food authentication.

According to statistics from 2011, approximately 18.5 million Muslims live in the European Union. In addition, it was the fastest-growing religion between 2001 and 2011, with the Muslim population increasing by 167%. As a consequence, there is an increasing demand for Halal food in Europe. With over 1.2 billion Muslims around the world, Halal food is of great significance, and there are restrictions involved in the production of such food. Not all food meets these requirements; this is where the concept of fraud or food adulteration arises. Noncompliance with these food products could result in not only religious issues but also serious legal and financial penalties. The fundamental reason for creating Halal food rules is to safeguard the health of Muslims by promoting the consumption of pure food and beverages and to provide assurance that food has been produced free from deceptive practices. Currently, several factors challenging the Halal food industry include food safety concerns, lack of trust, regulatory issues, and lack of standard implementation improvements in the Halal food supply chain. Ensuring the integrity of Halal food is becoming more important and urgent as a result of the increasing population of believers.

Several countries have adopted strict Halal laws. Taking these laws into account results in extra cost and time for Halal food. In particular, among the authentication agencies, there are pressures both internally and externally to have rapid, simple, reliable tests that can confirm both parties' and customers' confidence in products.

For these reasons, the possibility of quantifying a model to authenticate a finished product by using FTIR coupled with chemometrics application and external validation was provided using FTIR. The aim is to protect both Halal consumers and the religion from any possible negative issues by offering quick, reliable food product assessment. Furthermore, the proposed approach is simple and nondestructive to developing Halal food authentication.

Authentication of Halal Products Using FTIR Spectroscopy aims to create a multidisciplinary holistic overview of classification methods for Halal food authentication and detection of food adulteration. In addition, the book shows that Halal products can be chemically distinguished from non-Halal products, thus providing a strong and novel foundation essential for enhancing global trade and food quality, as well as combating food adulteration for hedonic and health reasons of different societies. The book offers an ample investigation of the chemometrics without overwhelming the reader with complex mathematics.

Jamshoro, Pakistan	Sarfaraz Ahmed Mahesar
Jamshoro, Pakistan	Syed Tufail Hussain Shah
Makkah, Saudi Arabia	Mohamed Fawzy Ramadan
Jamshoro, Pakistan	Waris Nawab Panhwar

Contents

1	**Introduction to Halal Food Authentication**		1
	1.1 Halal		1
		1.1.1 Introduction	1
		1.1.2 Halal Zabiha	5
		1.1.3 Electric Stunning	7
	1.2 Awareness of Halal Foods and Products		9
		1.2.1 Safety and Hygiene Related to Halal Food	10
		1.2.2 Relationship Between Halal, Hygienic, and Food Safety	10
		1.2.3 Halal Awareness	11
		1.2.4 Exposure to Awareness	12
		1.2.5 Halal Food for a Healthy Life	12
	1.3 Halal Food Logistics		13
		1.3.1 Halal Supply Chain	13
		1.3.2 Halal Food Ingredients	13
	1.4 Halal Pharmaceuticals		15
	References		17
2	**Techniques for Halal Authentication**		19
	2.1 Halal Authentication		19
	2.2 Chromatography for Halal Authentication		21
	2.3 Mass Spectroscopy		25
	2.4 ELISA		27
	2.5 DNA-Based Techniques		28
	2.6 Electronic Nose		30
	2.7 Sensors		31
	References		36
3	**Halal Food Authentication Using Chemometrics**		39
	3.1 Chemometrics		39
		3.1.1 History of Chemometrics	39
		3.1.2 Chemometrics in Scientific Research	40

	3.2	Chemometrics in Food Authentication	40
		3.2.1 Univariate Analysis	41
		3.2.2 Multivariate Analysis	43
		3.2.3 Machine Learning and Artificial Intelligence	48
	References		50
4	**Halal Food Authentication Using FTIR Spectroscopy**		**55**
	4.1	Introduction	55
	4.2	Importance of Halal Certification	56
	4.3	Foods That Need Halal Authentication	56
		4.3.1 Meat and Meat Products	58
		4.3.2 Pork Substitution	58
		4.3.3 Sausages	60
		4.3.4 Casings	60
		4.3.5 Blood Plasma	60
		4.3.6 Non-meat Ingredients	61
		4.3.7 Beverages or Drinks	62
	4.4	Perception of Food Fraud and Adulteration	62
	4.5	Challenges in Halal Food Authentication	63
	4.6	Spectroscopic Techniques for Food Authentication	64
		4.6.1 Theoretical Background of Infrared Spectroscopy	64
		4.6.2 Electromagnetic Radiation	65
		4.6.3 Infrared Absorption	65
		4.6.4 Normal Modes of Vibration	66
		4.6.5 Sources and Detectors	67
		4.6.6 Advantages of Fourier-Transformation	68
		4.6.7 Computers and Spectra	69
		4.6.8 Transmission Methods	70
		4.6.9 Reflectance Methods	74
		4.6.10 Attenuated Total Reflectance Spectroscopy	74
		4.6.11 Applications of Transmission and ATR Spectroscopy	75
		4.6.12 FTIR Spectroscopy and Chemometrics	76
	4.7	FTIR Spectroscopy for Halal Authentication	78
		4.7.1 Meat and Meat-Based Products	78
		4.7.2 Meatball Food Products	85
		4.7.3 Sausages	97
		4.7.4 "Rambak" Cracker	103
		4.7.5 Beef Jerky (Dendeng)	106
		4.7.6 Body Fats of Various Animals	106
		4.7.7 Oil and Fats	108
		4.7.8 Fish Oil	115
		4.7.9 Gelatin and Gelatin-Based Products	121
		4.7.10 Confectionary and Bakery Products	126
		4.7.11 Dairy Products	129
		4.7.12 Cosmetics	138
		4.7.13 Beverages and Drinks/Alcohol	141

	4.8	Portable and Hand-Held FTIR Spectroscopy to Food Control	145
	4.9	Current Status, Challenges, and Future Perspectives	150
	References		152
5	**Halal Meat, Fat, Oil, and Cosmetics Authentication**		**165**
	5.1	Meat Authentication	165
		5.1.1 Meat Geography Authentication	165
		5.1.2 Meat Origin Authentication	166
		5.1.3 Organic Versus Conventional Meat	167
		5.1.4 Identification of Meat Substitution	168
		5.1.5 Vegetable Protein and Animal Protein	168
	5.2	Oil and Fat Authentication	169
		5.2.1 Animal and Vegetable Fat	169
		5.2.2 Lard and Pork Authentication	170
	5.3	Cosmetics Authentication	171
		5.3.1 Halal (Permitted) Cosmetic Ingredients	172
		5.3.2 Critical Cosmetic Ingredients	173
		5.3.3 Detection of Haram Ingredients in Cosmetics	173
	References		175

Conclusion 181

Index 183

About the Authors

Sarfaraz Ahmed Mahesar is an Associate Professor at the National Center of Excellence in Analytical Chemistry (NCEAC), University of Sindh, Jamshoro, Pakistan. He earned his Ph.D. in Analytical Chemistry from the same institution in 2011. As a recipient of the IRSIP scholarship from the Higher Education Commission (HEC) of Pakistan, he completed part of his Ph.D. at Università di Bologna, Italy. In 2012, he was invited as a visiting researcher under the TÜBİTAK Research Fellowship Program at Middle East Technical University, Turkey. He has led several research initiatives, securing three projects under HEC Pakistan's NRPU program (2016–2017) as Principal Investigator (PKR 6.3 million and 1.5 million) and Co-Principal Investigator (PKR 5.8 million). His contributions to research and innovation include two patents and an extensive body of work comprising approximately 110 research and review articles and 12 book chapters in prestigious international journals and books. He has actively presented his research at national and international conferences.

An accomplished mentor, he has supervised 40 research scholars, including 10 Ph.D. and 30 M.Phil. students. His contributions to science have been recognized with the Research Productivity Award (RPA) by the Pakistan Council for Science and Technology (PCST) from 2010 to 2012. He holds key editorial roles, serving as an Academic Editor for the *Journal of Spectroscopy* and *Grain & Oil Science and Technology*, as well as the Managing Editor of the *Pakistan Journal of Analytical & Environmental Chemistry*. His dedication to advancing analytical chemistry and food science continues to shape the field through research, mentorship, and scholarly contributions.

Syed Tufail Hussain Sherazi is Professor of Analytical Chemistry at the National Centre of Excellence in Analytical Chemistry, University of Sindh, Jamshoro, Pakistan. He obtained his Ph.D. in Chemistry from the University of Sindh, Jamshoro (1997). He carried out his postdoctoral research at McGill University, Canada (2006), where he was also appointed as a visiting professor (teaching NMR spectroscopy). He presented his research work at more than 140 international and national conferences, as well as symposia and workshops. He is an author of more than 233 research articles published in peer-reviewed journals with a 559.2 impact factor, more than 8000 citations and 46 H-Index. He published one International US-PTO patent, and two patents were granted by the IPO, Pakistan Patent Office.

He has supervised 29 Ph.D. and 32 M.Phil. research scholars. He received the Research Productivity Award (RPA) from the Pakistan Council of Science and Technology (PCST), Ministry of Science and Technology, Pakistan, from 2009 to 2016. He received the Best University Teacher Award (BUTA) in 2013 from the Higher Education Commission (HEC), Pakistan, and the Gold Medal Award in Chemical and Pharmaceutical Sciences in 2016 by the Pakistan Academy of Sciences on the basis of his best performance in teaching and research, guiding research students, co-curricular activities, and participation in international conferences, seminars, and symposia as an invited speaker. On January 21, 2017, an award was given to Dr. Shah during the 2nd Pakistan Edible Oil Conference (PEOC) 2017 for his invaluable contribution to Edible Oil Processing Industries. He received the South Asia Triple Helix Association (SATHA) Award 2017 on the basis of a published patent, significant innovation, and effective contribution to society during the 2nd SATHA Invention to Innovation Summit 2017 of Sindh, held at Jinnah Women University, Karachi, on December 20, 2017. He received the Gold Medal Award from the University of Sindh Jamshoro during the Convocation held on March 9, 2018, for his valuable contribution to teaching and research. He received the

Best University Teacher/Researcher Award from the University of Sindh, Jamshoro, during the Convocation held on December 21, 2019, based on his contributions in terms of publications, impact factors, projects, national and international conference participation as well as teaching. He received the Academia Choice Award 2021 on World Teacher Day, October 5, 2021, at PC Hotel, Karachi, from the Provincial Minister of Education for his valuable contribution and efforts for the betterment of Education in Pakistan. He was successfully elected as a Fellow of the Pakistan Academy of Sciences (PAS) on December 7, 2024, by ratifying the recommendations of the PAS council.

Mohamed Fawzy Ramadan is Professor of Food Chemistry and Biochemistry at the Department of Clinical Nutrition, Faculty of Applied Medical Science, Umm Al-Qura University, Makkah, Saudi Arabia.

He obtained his Ph.D. (*Dr. rer. nat.*) in Food Chemistry from the Berlin University of Technology (Germany, 2004). He continued his postdoctoral research at ranked universities such as the University of Helsinki (Finland), Max Rubner Institute (Germany), Berlin University of Technology (Germany), and the University of Maryland (USA). In 2012, he was appointed as a Visiting Professor (100% teaching) in the School of Biomedicine, Far Eastern Federal University in Vladivostok, Russian Federation.

He has published over 450 research papers and reviews in international peer-reviewed journals. He has also edited and published several books and book chapters (Google Scholar *h*-index is 65, and there are more than 17,000 citations). He was an invited speaker at several international conferences. Since 2003, he has been a reviewer and editor in several highly cited international journals, such as *Earth Systems and Environment*, *Journal of Medicinal Food*, *eFood*, *Journal of Umm Al-Qura University for Applied Sciences*, and *Journal of Advanced Research*. He is the Editor-in-Chief of the *Journal of Umm Al-Qura University for Medical Science*.

He has received several prizes, including the Abdul Hamid Shoman Prize for Arab Researcher in Agricultural Sciences (2006), the Egyptian State Prize for Encouragement in Agricultural Sciences (2009), the

European Young Lipid Scientist Award (2009), the AU-TWAS Young Scientist National Awards (Egypt) in Basic Sciences, Technology and Innovation (2012), the TWAS-ARO Young Arab Scientist (YAS) Prize in Scientific and Technological Achievement (2013), and the Atta-ur-Rahman Prize in Chemistry (2014).

Waris Nawab Panhwar is currently a senior computer operator in the finance department of the Government of Sindh, Pakistan. He has a strong academic background in pharmaceutical sciences, analytical chemistry, and multidisciplinary research.

He completed his Doctor of Pharmacy (Pharm-D) in 2010 from the University of Sindh. He later pursued M.Phil. in Analytical Chemistry (2021) at the National Centre of Excellence in Analytical Chemistry, University of Sindh, Jamshoro. Currently, he is enrolled in a Ph.D. in Analytical Chemistry (since 2022) at the same institution, focusing on hydrogel-based research.

He has completed multiple courses in Machine Learning and Data Science and applies machine learning techniques in his research. He has authored a review paper on MXene and has a keen interest in multidisciplinary research, particularly in material science, nanotechnology, and biomedical applications.

Chapter 1
Introduction to Halal Food Authentication

1.1 Halal

1.1.1 Introduction

When discussing Halal food, we mainly refer to meat, fish, and animals. Halal meat is defined by two parameters: the method of slaughter and the type of animal. Halal and haram are clearly defined in the Hadith (sayings of Prophet Muhammad PBUH). Halal refers to something allowed for consumption, while haram denotes something prohibited. There is also the concept of Makrooh/Mashbooh (doubtful), which is optional for Muslims. Mashbooh refers to something uncertain due to differing perspectives among scholars or the inclusion of unknown components in culinary products. *Najis* refers to something unclean, while Makrooh refers to a disdain for a food product that is not explicitly haram but is regarded as undesirable by some Muslims (Khattak et al. 2011). Allah has given clear instructions for Halal and haram, and no one, under any condition, can change them. If something new arises that has not been previously discussed in Islam, it should be consulted with the Ulema (Islamic scholars).

In Arabic, the term Halal implies lawful or permissible. Non-Muslims generally identify the term "Halal" with only the foods that Muslims can consume. However, it encompasses all authorized actions and thoughts for Muslims. Halal affects all aspects of a Muslim's life, including clothing, work attitudes, relationships, child care, and business practices. Fellow Muslims must adhere to the Halal principle. Financial items, holidays, sports, films, and even how to play chess might be considered Halal or haram, respectively. Haram refers to everything that is prohibited for a Muslim. Haram is equally fundamental as the principles of Halal (Al-Teinaz et al. 2020).

Certain regions refer to "Zabiha" for animals slaughtered in the Islamic way. Food cooked by Muslims, Ahlekitab (Jews, Christians), and others also has different statuses in Islam. Table 1.1 lists animals that are prohibited for consumption in Islam.

To ascertain if fish and seafood are permissible, one must comprehend the guidelines of the various schools of Islamic jurisprudence and the regional variations in Muslim cultural traditions. While all Muslims generally consume fish with scales, certain sects forbid the consumption of catfish or other scaleless fish. Muslims hold diverse views on seafood, including mollusks and crabs. For instance, one must be aware of the requirements for exporting goods with seafood flavors in various parts of the world.

Milk and eggs are likewise Halal from Halal animals. In the West, cows are the leading milk producers, buffaloes are the leading producers of milk in Asia, and chickens are the leading producers of eggs. All other sources must be appropriately cited. Milk and eggs are used to make a wide range of goods. Cream, butter, and cheese are made from milk (Guyomarc'h et al. 2021). Most cheeses are produced using various enzymes, some of which may be considered Halal if derived from animals slaughtered in a Halal manner or from microorganisms. Enzymes used in food processing should be generated from microorganisms via fermentation, with raw materials and manufacturing methods sourced from non-haram or questionable sources (Ermis 2017). These enzymes might be deemed haram if derived from porcine sources or obtained from non-Halal-slaughtered animals. Emulsifiers, mold inhibitors, and other functional chemicals of unidentified origins can also raise concerns when consuming milk and egg products.

Except for alcoholic beverages and other intoxicants, plants and vegetables are typically considered Halal ingredients. However, in today's processing facilities, there is a possibility for vegetables and meats to be processed in the same facility and on the same machinery, increasing the risk of cross-contamination. The

Table 1.1 Lists of haram animals (Riaz and Chaudry 2003)

No.	Prohibited food
1	Carrion or dead animals
2	Flowing or congealed blood
3	Swine and its by-products
4	Animals slaughtered without pronouncing the name of God
5	Animals killed in a manner that prevents full drainage of blood
6	Animals slaughtered while pronouncing a name other than God
7	Intoxicants, including alcohol and drugs
8	Carnivorous animals with fangs
9	Birds of prey
10	Certain land animals, such as frogs or snakes

products become dubious since they might process vegetables with some animal-derived functional elements. To ensure that meals of plant origin remain Halal, processing aids and production techniques must be constantly supervised.

Halal is not only applicable to food products; it also applies to medicinal and cosmetic items. Halal practices extend beyond the manufacturing stage because a product can easily lose its certification if it becomes contaminated during transportation or storage before the sale (Zailani et al. 2017).

Numerous Muslims reside in nations including India, Australia, China, the European Union, the United States, and Canada, where Islam is not practiced as the official religion. As a result, the Muslim community contributes a large market segment to the contemporary food industry. However, the Muslim consumer group has received some disregard from the food industry. However, the Halal goods market has recently grown (Khan and Haleem 2016). Food product verification has been a major concern since ancient times, not only by consumers but also by producers and regulators. Halal authentication is the process of verifying whether a product comes into the domain of Halal or not.

There are several Halal authentication bodies throughout the world. Figure 1.1 shows the world bodies for Halal authentication and standardization.

Muslims are permitted to eat only the flesh of Halal animals. To be considered Halal, an animal must belong to a Halal species. A sane adult Muslim is required to slaughter the animal while uttering the name of God. The throat must be cut with a sharp knife in a way that causes complete blood loss and prompt demise.

Additionally, there are disparities in how Muslims view seafood, particularly mollusks and crustaceans like shrimp, lobster, and crab. The specifications and limitations apply not only to fish, seafood, and components made from such goods.

Cross-contamination of Halal products can occur during processing, packing, storage, and transportation, compromising their Halal designation (Supian 2018). Contamination of Halal in processing could be prevented by employing correct cleaning techniques and separating Halal output from non-Halal production. Functional components derived from animal sources, such as antifoams, must also be avoided when processing vegetables. Products made from plants and vegetables that intentionally include haram elements may be considered haram. It is clear that to keep vegetable products Halal, processing aids and production techniques must be closely inspected.

One of the primary concerns is the composition of food. As previously stated, vegetable products are Halal unless they have been tainted with haram ingredients or contain intoxicants. The guidelines for animal slaughter and the varieties of seafood acceptable for ingestion have already been covered. Here, we discuss some frequently utilized substances, including gelatin, glycerin, emulsifiers, enzymes, alcohol, animal fat and protein, tastes, and flavorings. Many manufacturers must have their facilities audited and their goods certified as Halal because most products fall into questionable categories.

Alcohol consumption is not permitted for Muslims, even in moderation. Ethanol's Halal status in the food industry is highly debated, but depending on its source and concentration, ethanol produced through natural fermentation at less than 1% is

Fig. 1.1 Examples of major Halal authentication bodies in different continents

considered a preserving agent and permitted. In contrast, ethanol produced through anaerobic fermentation at 1 to 15% is considered haram (Alzeer and Abou Hadeed 2016). Wine and beer should not be cooked with or used in other items as flavors because they contain alcohol. A Halal product becomes haram if even a small amount of alcohol is added to it. Both in the West and in China, cooking with wine, beer, and other alcoholic beverages is quite prevalent. Rice wine is a typical element in many recipes in Chinese cuisine. Chefs and product creators should refrain from using alcohol when creating Halal products.

Since alcohol is present in every biological system, fresh fruits contain traces of it. Alcohol may concentrate while the essences are being extracted from the fruits.

1.1.2 Halal Zabiha

There are two primary problems concerning animal welfare in the context of Halal slaughter without stunning. The first concern revolves around the welfare implications associated with the throat cut, while the second concern pertains to how the living animal is held and handled during the process. Certain countries employ highly stressful means of restraint to immobilize animals, such as the use of techniques like shackling and hoisting or shackling and dragging. Suspending calves by their rear legs induced a higher stress level than being held in an upright position. According to the guidelines provided by the World Organization for Animal Health (OIE) or the Office International des Epizooties (2007), it is advised that techniques of confinement that inflict pain and induce stress should be avoided when handling conscious animals.

Examining the existing body of literature concerning the physiological stress levels observed in cattle and sheep during on-farm and pre-slaughter handling reveals that cortisol levels in both scenarios are within a similar range. A further measure of pre-slaughter stress in cattle is the occurrence of vocalization, specifically moos and bellows, observed while forcing the animals into a restraint device or while they are being held within such a device. The occurrence of heightened vocalization in cattle during periods of constraint positively correlated with elevated cortisol levels. In a proficiently managed slaughter facility, the proportion of calves that emit vocalizations while confined in a restraint apparatus for religious slaughter is 5% or below. The mitigation of pre-slaughter stress has been found to positively impact the quality of meat and the well-being of animals.

Individuals effectively oversee the entities they quantify. Maintaining good standards necessitates continuously measuring and assessing handling and slaughter practices. Individuals tend to revert to previous negative behaviors unless they undergo ongoing evaluation. The practice of Halal slaughter without stunning can maintain an acceptable level of animal welfare, albeit with the need for increased attention to detail during the procedure compared to traditional slaughter methods using pre-slaughter stunning. Cattle provide a more significant welfare risk compared to sheep or goats.

This phenomenon can be attributed to the larger size of cattle, which presents challenges in terms of restraint and the duration of sensibility loss following slaughter without stunning, in contrast to sheep (Al-Teinaz et al. 2020). Humane slaughtering practices are important in the meat industry. Strict rules and regulations have been introduced to ensure the safety, hygiene, and welfare of animals used for food, focusing on achieving customer satisfaction through quality and safe products. The issue of religious slaughter remains critical, especially in the Muslim world, with little disagreement on the matter, including the allowance of pre-stunning before slaughter. Some Muslims object to stunning, stating that it is inconsistent with animal welfare principles and can lead to less blood drainage after stunning. There is a need to consider religious values and scientific knowledge to promote both economic value and food safety. The principle of stunning involves rendering an animal unconscious to minimize its pain during slaughter.

The standards set by the Standardization Expert Group of the Organization of the Islamic Conference (OIC) can be succinctly summarized as follows in Fig. 1.2:

(a) The animal designated for slaughter must adhere to Halal requirements, such as cattle, camel, sheep, etc., while excluding any unlawful (haram) animals, such as pigs or dogs.
(b) The animal intended for slaughter must be living or considered to be alive at the moment of the slaughtering process. The slaughtering process needs to be conducted in a manner that avoids inflicting unnecessary suffering onto animals and prioritizes the welfare and rights of animals.
(c) The individual responsible for performing the slaughter must adhere to the Islamic faith, possess good mental fitness, and possess a comprehensive understanding of the essential principles and requirements of animal slaughter.

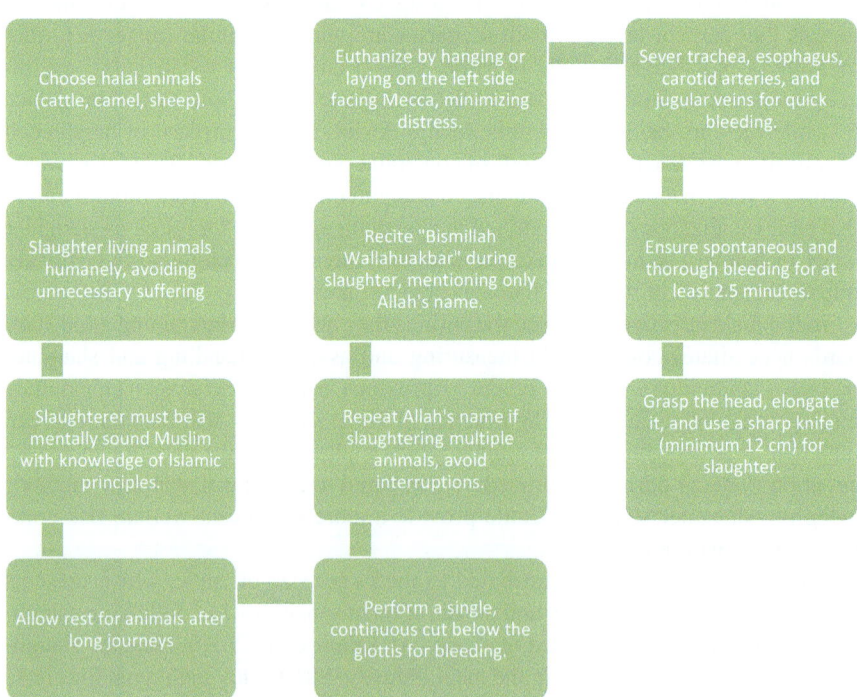

Fig. 1.2 Steps for Halal slaughter according to OIC

(d) If animals have traveled long distances, it is advisable to provide them with rest before their slaughter.
(e) The animal can be euthanized following the practice of hanging or laying it down, ideally on its left side, with its face directed towards Kiblah, which is the direction of Makkah Al-Mukaramah. Efforts should be made to minimize the distress experienced by the animal throughout the process of hanging or laying, and it is imperative to avoid prolonged periods of suspension or confinement in such positions.
(f) During the process of animal slaughter, it is customary for the slaughterer to recite the phrase "Bismillah Wallahuakbar," which translates to "In the Name of Allah and Allah is the Greatest." The slaughterer needs to refrain from mentioning any name other than Allah, as doing so would render the meat non-Halal.
(g) The repetition of mentioning the name of Allah is necessary during the killing of multiple animals or a continuous group of animals. If the constant process is interrupted, invoking Allah's name again is essential. Slaughtering should be limited to a single occurrence for each animal. Slaughtering, sometimes called "sawing action," is permissible for the duration necessary to complete the slaughter. It is important to note that the knife should not be lifted from the animal during the process.
(h) The process of Halal slaughter involves initiating an incision in the neck, specifically below the glottis (also known as Adam's apple), and extending it beyond the glottis for animals with lengthy necks.
(i) The act of slaughter involves the severing of the trachea (halqum) and oesophagus (mari), as well as both the carotid arteries and jugular veins (wadajain) in order to expedite the process of bleeding and subsequent death of the animals.
(j) The occurrence of bleeding will be both spontaneous and thorough. In order to achieve complete bleeding, the duration of bleeding should not be less than 2.5 minutes.
(k) The one responsible for the act of slaughtering should firmly grasp the head with their left hand, exerting pressure to elongate it, while employing a sharp slaughtering knife carried in their right hand to incise the throat. The length of the sharp edge of the knife utilized for slaughter should be a minimum of 12 cm.

1.1.3 Electric Stunning

Electric stunning, also known as electronarcosis, was developed in the 1920s and is used in many countries. However, its use is not always allowed in Muslim countries due to religious concerns. In some countries, reversible types of percussive and electric stunning are used prior to Halal slaughter. The most common method of stunning chickens is electric stunning, which causes cardiac arrest and electrocution. The conventional method of slaughtering involves stunning the animals before exsanguination. Religious slaughter, on the other hand, consists of a throat cut without stunning to bring about a quick death. The current, resistance, and voltage-current relationship are considered in electric stunning to ensure a minimum current is used. CO_2 stunning consists of exposing animals to a high concentration of CO_2, which leads to the loss of consciousness.

The Jewish religion has strict dietary laws known as kosher that regulate food selection, slaughtering, processing, and consumption. The laws mainly focus on three aspects: the type of animal that is allowed, the prevention of blood, and the mixture of meat and milk. A skilled religious person does slaughtering called a "shochet" who uses a specific knife called "chalef" to perform the slaughter according to Jewish laws. The knife must be extremely sharp with a blade diameter equal to twice the width of the animal's neck. The animal is not to be stunned before slaughter, and the shochet must bless the animal and check the slaughter for accuracy. After slaughtering, the animal is inspected by a rabbinically skilled assessor for any defects. The rules of kosher do not allow trimming of any defective organs, but there may be religious attention given to the animal if it is estimated to pass away within a year. US kosher standards mainly focus on examining lungs, and other organs are checked if a potential problem is detected. There are also rules regarding the salting of kosher meat, where only the surface of the meat is salted, and the salt penetrates to about half a centimeter.

The definition of the word "dhabh" is "to slaughter." Islam requires that certain conditions be met for *dhabh* to be considered Halal. Generally, the animal must come from a species that is recognized as Halal and must be killed by a Muslim in the name of Allah, first severing the jugular vein with a sharp knife and allowing total blood drainage, which causes the animal to die quickly (Arshad et al. 2023).

The Halal slaughter method induces rapid loss of consciousness in animals through the anemia of the brain, which is brought about by the simultaneous and instantaneous severance of the carotid arteries with a sharp knife. This implies that Halal slaughter does not elicit any pain experience.

The idea was empirically validated by a research team at the University of Hannover in Germany, who employed electrocardiography (ECG) and electroencephalography (EEG) data to compare Halal and shocking slaughtering methods. The utilization of the Halal technique of slaughter resulted in the absence of any discernible alteration in the EEG graph during the initial 3-second period after the incision, suggesting that the animal did not experience any sleep-like state of unconsciousness induced by the removal of substantial volumes of blood from its body. Subsequently, the electroencephalogram (EEG) registered a null measurement, signifying the absence of pain. However, during this period, the heart continued to beat, and the body exhibited intense convulsions as a reflexive response originating from the spinal cord. The abovementioned stage is particularly disagreeable for observers, who mistakenly believe that the animal experiences suffering, although, in reality, its brain no longer registers any sensory signals of pain originating from the incision. The state of profoundness marked the subsequent 3-day interval (Abd El-Rahim 2020).

The utilization of various slaughtering techniques, such as stunning, closing, and electrical shock, commonly employed in Western countries, can impede the proper execution of the bleeding process. Blood is a nutrient-rich medium supporting the proliferation and development of diverse microorganisms. Consequently, consuming shocked meat can potentially introduce bacterial infections to individuals. Furthermore, the potential contamination of organs and muscles with brain tissue, resulting from specific methods of stunning, can serve as a potential reservoir for infection with prior illnesses that are currently incurable.

Research has demonstrated that stunning can expedite meat aging and alter quality attributes, such as moisture depletion and coloration. The utilization of contemporary slaughtering techniques has been found to have a detrimental impact on the glycolysis process, also known as rigor mortis, leading to the production of inferior-quality meat.

The Halal way of slaughtering has been empirically verified as the most effective technique for ensuring optimal bleeding. In addition, the Halal method holds considerable significance in human health, as it safeguards customers against the spread of infectious diseases and plays a substantial role in ensuring the safety and hygiene of meat products.

It is advisable that individuals who do not identify as Muslims also consider abstaining from utilizing contemporary methods of animal slaughter and instead adopt the non-stunning and hand-slaughter Halal approach to avail themselves of the associated benefits.

1.2 Awareness of Halal Foods and Products

Awareness gives the primary material for developing subjective ideas about one's encounter with something (Abdullah 2006). Awareness of Halal food is growing globally due to readily available digital information. Halal awareness and product ingredients have a substantial impact on Muslims' intentions to purchase Halal packaged food from non-Muslim producers, with religious belief, exposure, and certification/logo all potentially contributing to awareness (Azam 2016). Due to its acceptance as a substitute standard for safety, hygienic practices, and quality assurance of the foods and beverages we regularly consume or drink, Halal products or foods are currently attracting widespread discussion. As a result, both Muslim and non-Muslim consumers generally accept goods or foods that are prepared following Halal regulations.

Halal foods and beverages are produced strictly following the Holistic Halal Assurance Management System. For a Muslim consumer, the product has complied with Sharia law requirements. Religious beliefs, subjective norms, and Halal awareness all have a substantial impact on young Muslim consumers' intents to purchase Halal food (Nurhidayana and Juniarti 2020). For a non-Muslim consumer, it symbolizes hygiene, quality, and product safety. As a result, customers nowadays are apprehensive and conscious of everything they use, consume, and ingest (Ambali and Bakar 2014).

1.2.1 Safety and Hygiene Related to Halal Food

Halal focuses on hygiene, covering various topics related to physical appearance, clothing, tools, and workspace where food, beverages, and other goods are processed or manufactured. The goal is to ensure that every food product is healthy, hygienic, and safe for human consumption.

Hygienic food, beverages, and products lack *najis*, contamination, and dangerous microorganisms in Halal. Therefore, it is clear that Halal is highly specific about food, particularly in keeping our surroundings and ourselves clean to prevent infections. A safe meal, drink, or product is one that, when made, consumed, or used as intended, does not harm its users, whether they be Muslims or not. Halal food enterprises can attract non-Muslim consumers as a target market since they prioritize sanitation, cleanliness, and food quality over religious values (Mathew 2014).

The makers should make the required efforts to comply with Good Manufacturing Practices (GMP) and Good Hygiene Practices to ensure our safety. In order to ensure that the products are regularly made to their specifications and the Halal prescriptions provided by the Halal Certification Body, producers must implement a mix of production and quality control methods known as good manufacturing practices.

1.2.2 Relationship Between Halal, Hygienic, and Food Safety

Consuming Halal food as directed by Allah (SWT) must be seen from a broader perspective. From the standpoint of the quality and overall goodness of what we should eat, drink, and use daily, the notion of Halal fully includes all facets of human life. Therefore, food must be Halal, high-quality, safe, and hygienic for Muslims. It demonstrates that all food is Halal except those expressly listed as haram in the Qur'an.

Islam only allows its adherents to consume legal, hygienic, secure, and healthy foods, beverages, and products that adhere to Shariah and the Holy Quran. Consuming Halal is, therefore, not only necessary for serving Allah but also proves that the ingredients and materials are not damaging to one's health because Allah only gave His permission for things that are beneficial for human existence.

Through Halal, Islam highly emphasizes cleanliness, safety, and hygiene. It covers every aspect of personal hygiene, attire, and facilities used in the production or preparation of food. Cleanliness and health are the foundation of Halal. The goal is to guarantee that the foods, beverages, and items people consume or use are safe and do not affect human health. It is important to remember that using and consuming Halal items is required in Islam to serve Allah (SWT). Muslim communities must be aware of the components of food and beverages and the handling and packaging procedures. Processed meals, beverages, and goods are only Halal if the raw materials and ingredients are Halal and are used in full compliance with Islamic laws. Muslims must, therefore, be conscious of the Halal aspects of the food they consume.

1.2.3 Halal Awareness

The definition of "awareness" is the knowledge or comprehension of a specific topic or circumstance. In the context of Halal, the word "awareness" literally means having a particular interest in or experience with something and/or being knowledgeable about what is happening with Halal foods, drinks, and products. As a result, awareness refers to a person's perception of and cognitive response to the state of the food, drink, and other substances they use.

A person may be acutely aware of a problem relating to a Halal component permitted by Allah, or they may be only partially aware, subconsciously aware, or both. It may be concentrated on an internal condition, such as visceral emotion, or things outside of oneself through sensory perception. Awareness offers the foundation for creating arbitrary theories about how one's experiences relate to anything. Therefore, being aware of something is a fundamental aspect of being human. Self-awareness is the priority. To be aware is to have one's private views about how something is right now. As a result, various people have varying knowledge about a subject.

To put it another way, raising people's knowledge regarding hazards associated with anything that could harm human life and what they can do to lessen their exposure to it is what awareness means. Therefore, awareness in the context of Halal can be seen as the increasing understanding of what Muslims should consume, use, and possess.

A group of individuals use their religious ideas and rituals to interpret and react to what they consider supernatural and sacred. The majority of religions encourage or forbid specific behaviors, including consuming. Islam makes it very clear that whereas non-Halal products are prohibited for ingestion by humans, Halal foods, drinks, and other items are permitted.

Assert that the identity, orientation, knowledge, and beliefs of members of various religious groups impact how they choose to spend their money. It demonstrates that awareness of consumption behavior can come from sources such as religion and belief. Religious knowledge or belief is one of the key determinants of dietary avoidance, taboos, and specific regulations, particularly regarding meat.

The best way to decide what to eat is to use your religious beliefs or knowledge because different religions have different food taboos. For example, pig and meat that has not been ritually slain are forbidden in Judaism and Islam, while pork and beef are forbidden in Buddhism and Hinduism. Even though some religions' dietary restrictions may be reasonably stringent, many individuals adhere to them. More than 75% of Muslim immigrants to the United States adhere to their Islamic dietary requirements. It demonstrates that Muslims are still conscious of Halal because of their religious knowledge and belief, regardless of where they choose to live.

1.2.4 Exposure to Awareness

Living in a modern science and technology age leads to a wide range of culinary products. This evolution coincides with a growth in chemicals and ingredients to meet needs and perfect food production. The wide range of meals and items on the market frequently confounds consumers; most are oblivious to what they have ingested. It is believed that consumers must place their trust in the information source and the information they receive and rely on the vendor or outside observers. In order to provide consumers with guidance, it is crucial to educate them about food selection and expose them to it. Educational exposure is one of the best ways to educate people about the sorts of food they eat in the context of safety and hygienic conditions, which is Halal's primary goal. By educating them, we can help people become aware of what they consume every day and choose wisely. Delivering materials and education about food safety to various target populations is, therefore, a primary obligation of the government or agency in charge of Halal. Consumers, school-aged children, and workers in the food business all need to get food safety education. Consumers in Muslim nations might be introduced to Halal to raise their level of awareness through education and learning. In this era of technological growth, the government may inform the public about Halal using various media. Daily newspapers, television, radio, the internet, and other media can all be used to educate people. All of them could be crucial in conveying information on Halal alerts and exposure. Thus, teaching exposure can provide information about Halal concerning what Muslims eat.

1.2.5 Halal Food for a Healthy Life

Halal food consumption supports a healthy lifestyle and favors human development, including intelligence, *akhlak*, morality, and psychology (Sawari et al. 2015). Health concerns related to religious identity and the degree of acculturation in the foods we consume daily might also influence people's knowledge of Halal food or products for consumption. For people to remain healthy, it is crucial, for instance, to ensure that the meat comes from a healthy animal. Most bad health is caused by poor nutrition and the unhealthy foods consumers consume regularly. This is closely related to the case of Halal consumption because Allah's primary objective in Halal is to guarantee people a long and healthy life. Halal exhorts to maintain complete devotion to creating, delivering, and providing consumers with clean, safe meals, and products. In other words, people should see Halal items as a mark of excellence, cleanliness, and safety.

1.3 Halal Food Logistics

The Halal Supply Chain Model optimizes Halal food supply chains by focusing on transportation, warehousing, and terminal operations, reducing vulnerability to contamination (Tieman et al. 2012). Halal is not static; it evolves from being associated with Muslim businesses, Halal products, and supply chains to a Halal value chain. The logistics of Halal products are being questioned by both the food industry and the logistics sector, as Halal extends upstream and downstream throughout the supply chain. This has led to measures to certify logistics operations following Halal standards. The Halal sector relies on Halal logistics and supply chain management (SCM) to extend the Halal integrity from the point of production to the point of consumer purchase. Halal logistics performance, particularly for meat goods, can increase consumer loyalty in Muslim countries, with aspects such as supplier service quality, perceived service value, and customer happiness all playing a role (Masudin et al. 2018). However, there is a dearth of literature on Halal food logistics and SCM. The Halal industry may benefit from this Halal Supply Chain Orchestrator's (HSCO) ability to authenticate Halal products, access new markets, and reduce costs. Hazard Analysis and Critical Control Point (HACCP) is a methodical approach to analyzing potential risks in a food supply chain, determining the critical control points where the hazards may occur, and selecting which of points are essential for food safety.

1.3.1 Halal Supply Chain

The Halal supply chain aims to satisfy the needs and requirements of both Halal and non-Halal customers by managing Halal products from various suppliers to various end customers. This process involves numerous parties located in various locations. Halal supply chains differ from regular supply chains in that their primary objective is to ensure that the product's Halal status is maintained throughout the entire supply chain process and to ensure consumer satisfaction. It also considers the dedicated cold chain, including warehousing (Khan and Haleem 2016).

1.3.2 Halal Food Ingredients

The Halal industry is not only about Halal meat; it also includes Halal food, pharmaceuticals, cosmetics, lifestyle, and even Halal services (Elasrag 2016). Humans have utilized food additives for ages, indicating that they are not a recent development. Historically, our predecessors employed salt to preserve meats and fish while enhancing the taste of various dishes by adding herbs and spices. Additionally, they utilized sugar to preserve fruits and used a vinegar solution to pickle olives and cucumbers.

In contemporary times, the proliferation of processed foods has led to a significant surge in the incorporation of additives, resulting in the chemical adulteration of food products. There has been considerable debate surrounding the potential risks and advantages of food additives.

Food processing refers to a collection of methodologies and strategies employed to convert unprocessed resources into consumable food products or modify existing food items into other forms suitable for human or animal consumption, whether in domestic settings or within the food industry.

The optimal time to consume homemade food is typically immediately after preparation. The transportation and storage of food on a big scale, necessary to meet the demands of supermarkets and other food retailers, is a crucial aspect of the food supply chain. Home-cooked food must maintain its quality and freshness for an extended duration compared to different types of cuisine.

Food additives are deliberate additions of ingredients to food products to fulfill specific technological purposes, such as enhancing color, sweetness, or preservation. Additives are of such significance that they are employed in certain organic food products as well.

In numerous nations, a significant amount of food is wasted due to microbial proliferation, leading to spoilage before consumption. Foodborne illness exemplifies the hazards associated with consuming contaminated food, and it is plausible that the absence of preservatives might result in a higher prevalence of such cases.

A food additive can be described as a natural or artificial substance, distinct from the primary raw ingredients, utilized while manufacturing a food product to enhance its overall quality (Fig. 1.3). Additionally, it encompasses any substance that has the potential to influence the properties of any food, including those employed in the various stages of food production, processing, treatment, packaging, transportation, or storage.

The sourcing of components for Halal food processing must adhere to Halal standards, while the processing must comply with Islamic principles and regulations. It is imperative to thoroughly examine final compositions to ascertain the presence of any alcoholic substances used in the processing stage. In the ultimate formulation, the presence of alcoholic constituents must adhere to the prescribed thresholds. Packaging materials must be devoid of any elements that are considered haram. Preventing cross-contamination with non-Halal substances is crucial. Ensure that equipment is meticulously cleansed with detergents that are deemed acceptable. In food processing, thoroughly clean all equipment after manufacturing non-Halal ingredients, mainly when the same equipment is used for processing Halal and haram foods. To prevent cross-contamination, segregate all Halal products during the storage process.

The unresolved issue of ingredient and product conformity in the food and pharmaceutical industries persists when seen through the lens of Halal standards. One of the challenges faced by the Islamic community involves the establishment of a food laboratory dedicated to the analysis of components, with the subsequent

1.4 Halal Pharmaceuticals

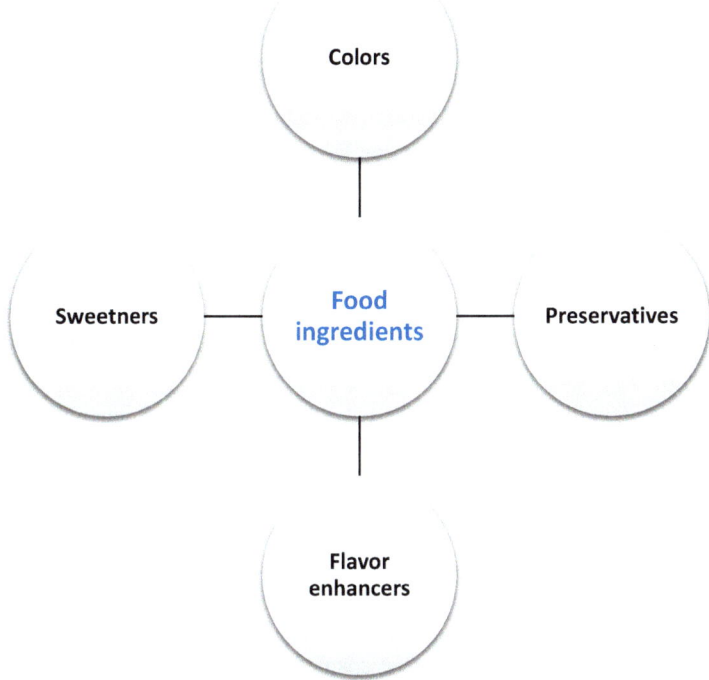

Fig. 1.3 Food ingredients used as additives

dissemination of findings to the Muslim community. Ensuring the origin and processing of food additives align with Halal regulations while abstaining from alcohol-based transporters is important (Ruževičius 2012).

1.4 Halal Pharmaceuticals

The 2.4 billion Muslims worldwide who purchase goods are becoming more aware of and interested in Halal medicines and cosmetics. Until 2024, the global Halal market is projected to develop at a compound annual growth rate of 6.8%. The Halal pharmaceutical sector is confronted with difficulties in creating a good accreditation and Halal management system, but the expanding global Muslim population provides the potential for expansion (Norazmi and Lim 2015). Halal pharmaceutical makers should emphasize Halal ideals and create a broader picture of Islamic values to reach a larger market segment and encourage sustainable practices (Mohezar et al. 2017). Halal cosmetics have a greater market appeal among non-Muslim consumers, who associate these products with ethical consumerism and stricter quality assurance requirements, in addition to the substantial Muslim population that shows a high commitment to Halal products. Halal can also establish itself as a standard of acceptable quality and conformity in trading with Muslims for non-Muslims. Ingredients from pig, carrion, blood, human body parts, predatory animals, reptiles, insects, and others are prohibited in Halal cosmetic goods. For cosmetic materials to be certified Halal, they must have been killed following Islamic law. Halal cosmetic products must always be prepared, processed, manufactured, stored, and transported under clean and hygienic conditions. Emphasis is placed on the lack of filth. The purpose of Halal certification is similar to the objectives of most quality assurance processes (such as GMP and HACCP). As a result, consumers must regard Halal cosmetics with the Halal emblem as a sign of quality, purity, safety, and cleanliness. Although pharmaceutical products have advanced scientifically far more than cosmetic products, as shown by the many Halal-certified medicine items on the market, further research is needed to develop cosmetics as Halal products to meet worldwide demand.

Given that Islamic law mandates that all Muslims consume only Halal and healthy products, understanding the origin of raw materials and the production method of cosmetic ingredients is essential for Muslim customers. Furthermore, examining cosmetic items requires technical expertise regarding the ingredients, origins, and manufacturing processes. Complex cosmetic products comprise several highly processed substances that may come from plants or animals. Cosmetics can accidentally be consumed (like lipstick), breathed (like scents), or absorbed through the skin (like alcohol or substances of crucial origin) after application. Given the difficulties, producers must ensure that cosmetics are carefully created to be Halal in composition and to fully support daily Islamic rituals (such as *wuzu*, which involves washing one's hands before praying or reading the Qur'an). When creating Halal cosmetics, several obstacles must be considered. It is highly challenging to confirm that animal-based cosmetic ingredients, including gelatin, lecithin, glycerol, fatty acids, and collagen, are Halal. Some coloring compounds may be made from insects and are therefore considered haram.

Additionally, substances derived from cows provide another difficulty because the animals could be killed in a non-Halal way. To comply with Islamic ritual regulations, cosmetic products must be produced using Halal materials and performed effectively. In order for Muslims to follow their rituals as prescribed, for example, cosmetic materials applied to the skin must either be entirely rinsed off or be able to penetrate a lacquered nail to allow for enough rinsing. The production of Halal cosmetics and the evaluation of product performance are still in their infancy despite the existence of haram substance detection tools. Most cosmetics are produced by non-Halal cosmetic firms, whose production processes do not follow the guidelines of Halal science, underscoring the necessity of creating guidelines for such a purpose.

Additionally, there is a global shortage of guidelines for the creation of fully Halal cosmetics and assessment methods. Halal cosmetics are goods made with Halal materials and created in line with Halal principles designed to be applied to a particular part(s) of the body, whether as rinse-off or leave-on products, to enhance, protect, and modify the body's appearance. Water, oils, surfactants, polymers, organic solvents, colorants, proteins, vitamins, plant extracts, preservatives, and antioxidants are just a few elements that make up cosmetics. Cosmetic producers must carefully consider each ingredient and its source before developing and manufacturing due to the intricate combination of ingredients in a cosmetic product.

The origin of the materials used in creating and producing Halal cosmetics significantly impacts the final result and general effectiveness of the finished product. Manufacturers, not regulators, are responsible for demonstrating the safety of the ingredients used in Halal cosmetic goods.

A Muslim can be confident that a product is Halal certified if it does not contain pork, substances derived from pig, or other ingredients that are against Islamic law (such as alcohol or other products or derivatives of animals that are not authorized). Additionally, it assures the buyer that the product was produced using machinery designed explicitly for Halal pharmaceuticals. Furthermore, any animal products or derivatives used are anticipated to come from animals that were killed in conformity with Islamic law. Many new Halal Control (HC) items are being developed, including medications, additives, flavors, enzymes, and food supplements.

References

Abd El-Rahim IH (2020) Recent slaughter methods and their impact on authenticity and hygiene standards. In: The Halal food handbook, pp 81–91

Abdullah AN (2006) Perception and awareness among food manufacturers and marketers on Halal food in the Klang Valley. Unpublished Master Dissertation, Universiti Putra Malaysia, Selangor

Al-Teinaz YR et al (2020) The Halal food handbook. Wiley

Alzeer J, Abou Hadeed K (2016) Ethanol and its Halal status in food industries. Trends Food Sci Technol 58:14–20

Ambali AR, Bakar AN (2014) People's awareness on Halal foods and products: potential issues for policy-makers. Procedia Soc Behav Sci 121:3–25

Arshad MS, Khalid W, Noreen S (2023) Methods and technology used. In: Abdul Rahman NA, Mahfooz K, Hassan A (eds) Technologies and trends in the Halal industry, Chapter 8, 1 edn. Routledge

Azam A (2016) An empirical study on non-Muslim's packaged Halal food manufacturers: Saudi Arabian consumers' purchase intention. J Islam Mark 7(4):441–460

Elasrag H (2016) Halal industry: key challenges and opportunities

Ermis E (2017) Halal status of enzymes used in food industry. Trends Food Sci Technol 64:69–73

Guyomarc'h F et al (2021) Mixing milk, egg and plant resources to obtain safe and tasty foods with environmental and health benefits. Trends Food Sci Technol 108:119–132

Khan MI, Haleem A (2016) Understanding "Halal" and "Halal certification & accreditation system" – a brief review. Saudi J Bus Manag Stud 1(1):32–42

Khattak JZK et al (2011) Concept of Halal food and biotechnology. Adv J Food Sci Technol 3(5):385–389

Masudin I et al (2018) Halal logistics performance and customer loyalty: from the literature review to a conceptual framework. Int J Technol 9(5):1072–1084

Mathew VN (2014) Acceptance on Halal food among non-Muslim consumers. Procedia Soc Behav Sci 121:262–271

Mohezar S et al (2017) Tapping into the Halal pharmaceutical market: issues and challenges. In: Contemporary issues and development in the global Halal industry: selected papers from the international Halal conference 2014. Springer

Norazmi MN, Lim LS (2015) Halal pharmaceutical industry: opportunities and challenges. Trends Pharmacol Sci 36(8):496–497

Nurhidayana A, Juniarti RP (2020) Bagaimana Generasi Milenial Membeli Makanan Halal? Peran Religious Belief, Subjective Norm, dan Halal Awareness. Jurnal Manajemen Dan Bisnis Sriwijaya 18(4):213–224

Riaz MN, Chaudry MM (2003) Halal food production. CRC Press

Ruževičius J (2012) Products quality religious-ethnical requirements and certification. Ekonomika ir vadyba 17:761–767

Sawari SSM et al (2015) Evidence based review on the effect of Islamic dietary law towards human development. Mediter J Soc Sci 6(3 S2):136

Supian K (2018) Cross-contamination in processing, packaging, storage, and transport in Halal supply chain. In: Preparation and processing of religious and cultural foods. Elsevier, pp 309–321

Tieman M et al (2012) Principles in Halal supply chain management. J Islamic Mark 3(3):217–243

Zailani S, Iranmanesh M, Aziz AA, Kanapathy K (2017) Halal logistics opportunities and challenges. J Islam Mark 8(1):127–139. https://doi.org/10.1108/JIMA-04-2015-0028

Chapter 2
Techniques for Halal Authentication

2.1 Halal Authentication

Halal verification is a rigorous process that ensures that the requirements of Islamic law are followed when producing and consuming meat and meat products. It is composed of following Sharia/Islamic laws and Halal standards in food, abstaining from using haram ingredients or food sources.

Consumers' packaged goods should be prepared, stored, and transported in Halal manners defined in Sharia law. Halal authentication of consumer packaged goods faces challenges. In order to achieve this, establishing a standard approach, sound traceability system, and constant monitoring are needed for Halal authentication. In consumer packaged goods, Halal authentication goes beyond meat to certify the Halal status of numerous products. The process is comprehensive, addressing issues such as production processes and technological processing, identifying undeclared components, and preventing species replacement in Halal food items. Halal food verification is more than a technical process; it also requires adherence to Islamic law and ethical principles. Consumers play an essential role in this authentication process because it involves discerning and highly involved behavior. By actively selecting items that meet Halal standards, customers help to promote transparency and ethical practices in the food sector.

Analytical procedures are an important component of Halal food authentication, serving two purposes: assuring food safety and safeguarding consumers from fraud (Fig. 2.1).

By implementing these strategies, the sector can protect itself from infractions, supply consumers with real Halal products, and create trust in the marketplace through openness. Halal authentication is a comprehensive strategy that combines technological precision, ethical concerns, and consumer participation to ensure that food products are produced and consumed following Islamic norms (Wilson and

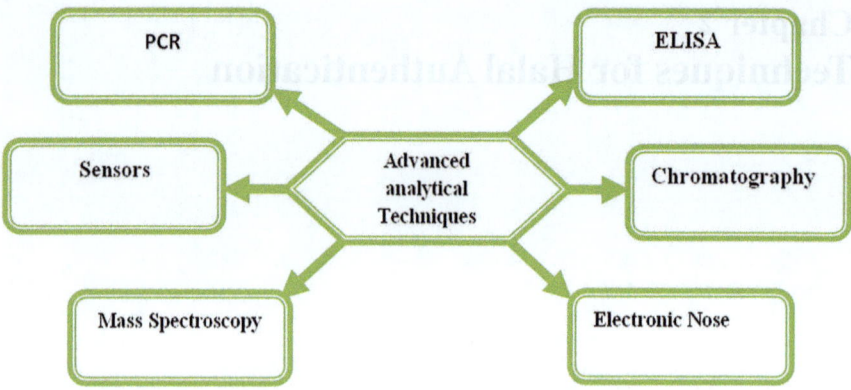

Fig. 2.1 Advanced analytical techniques used for Halal authentication

Table 2.1 Ingredients/components that are most commonly analyzed to check for Halal authentication

Ingredient/component	Islamic ruling	Analysis
Pork and lard	Haram (prohibited)	Identified through source analysis and labeling
Alcohol (liquor/*Khamr*)	Haram (prohibited)	Defined in Islamic law; detected through chemical analysis
Gelatin	Source-dependent	Analyzed through chemical and crystallization methods
Collagen	Typically haram	Often derived from swine, analyzed through source identification
Emulsifiers (mono/diglyceride)	Haram if from unlawful origin	Analyzed for fatty acid profiles to determine the origin
Fats and oils	Analyzed for origin	Identification through hydrolyzed fatty acid profiles
DNA	Used for fraud identification	Detects porcine, canine, or other contaminations

Liu 2011; Nakyinsige et al. 2012; Jaswir et al. 2017; Premanandh and Bin Salem 2017; Ng et al. 2022).

Confirming that a good or service is Halal, or legal under Islamic law, is known as Halal authentication (Table 2.1).

For Muslims who adhere to Islamic dietary regulations and want to make sure the things they consume are Halal, their verification is crucial. Halal certification, which comprises an independent evaluation, verification, and supervision of a business and its goods about compliance with Islamic dietary regulations, can be used to authenticate Halal products. The authentication of Halal food can be challenging due to the complex nature of food and the increasing number of food adulterants that cause detection difficulties (Ng et al. 2022).

2.2 Chromatography for Halal Authentication

Chromatography is an analytical process for separating molecules from complicated mixtures utilized in many industries, including pharmaceuticals, food, oil, and natural product synthesis (Kisley and Landes 2015). Gas chromatography (GC) and liquid chromatography (LC) are two strong techniques used in analytical chemistry to separate, identify, and quantify different chemicals in a sample.

GC is an analytical technique that analyzes, separates, and identifies volatile compounds in the mixture. In GC, the sample is injected in gas or liquid form into a gaseous mobile phase that carries the sample through the column containing a stationary phase in liquid or solid form. The sample partitions due to differential affinity towards the mobile and stationary phases.

The GC instrument comprises an injection port, column, gas flow equipment, oven, heater, and detector. GC is an efficient, sensitive, and selective technique that utilizes a small sample and provides good results.

LC is a technique used to separate components of a mixture in liquid form. The mobile phase used in LC is in liquid form, and the stationary phase is bound to the column of simple components separated according to their affinities towards the mobile and stationary phases. LC utilizes high pressure for efficient and fast separation. Liquid chromatography is hyphenated with mass spectroscopy to analyze and detect complex mixtures and their components.

HPLC and GC techniques are employed to check the authenticity and traceability of food items, enabling the identification of unique compounds and the determination of their origin. HPLC is combined with other methods to analyze the fatty acid composition of food products, while GC is utilized to analyze volatile compounds in food samples (Fig. 2.2).

When considering chemical types, GC is more suited for analyzing smaller, volatile, and thermally stable substances. On the other hand, LC is more resilient and appropriate for larger compounds that are less volatile or nonvolatile. Derivatization techniques are employed in LC to convert analytes into detectable derivatives using specific detectors. In GC, certain compounds are derivatized to increase their volatility, but this process increases the complexity of analysis and the analysis time. Care should be taken when derivatizing as it should not impact analysis.

Single-column or one-dimensional GC is simple, efficient, and effective for less complex samples. While dealing with high complexity, samples may prove ineffective, and additional columns are required called 2-D GC. Using one-dimensional GC for complex samples can result in unresolved analytes and peaks. 2D GC possesses greater capacity and more resolution in larger samples. 2D GC and 2D LC have many food, pharmaceutical, and environmental applications.

In 2D GC and 2D LC, columns are connected through a modulator. This allows for integrating analytical data acquired from the first column, such as retention times (tR), with the corresponding data from the second column. Consequently, a graphical representation of two retention times can be generated. The use of 2D GC

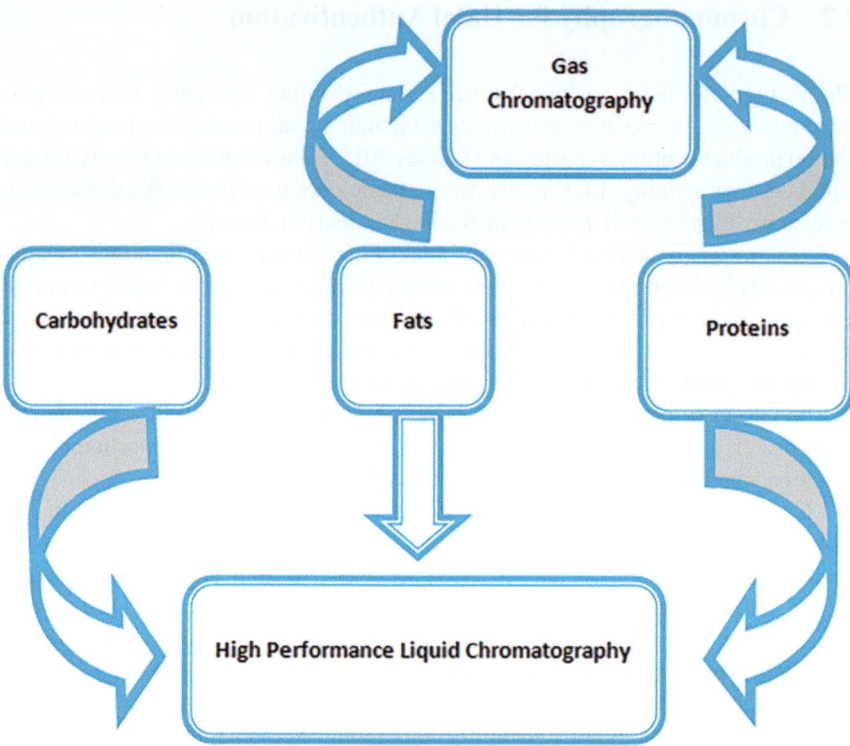

Fig. 2.2 Types of food components and chromatographic techniques used for their analysis

and 2D LC in conjunction with mass spectrometry (MS) is prevalent in separating various constituents inside intricate mixtures, particularly in metabolomics research, owing to its exceptional separation capabilities. The comprehensive two-dimensional GC x GC technique has been extensively used to analyze metabolites, namely, all forms of fatty acids in lard samples obtained from dietary sources. On the other hand, LC coupled with 2D LC x LC is commonly employed to analyze peptides. This analytical approach facilitates the identification of pig and porcine gelatines.

Chromatographic-based approaches have proven to be dependable for conducting Halal authentication analysis. Nevertheless, given the substantial volume of data involved, the utilization of chemometrics in handling large datasets becomes inevitable. Chemometrics refers to applying statistical and mathematical techniques to evaluate data objectively. This consists of obtaining pertinent and significant information from chemical measurements, regardless of whether the responses are directly or indirectly related. Chemometrics, also known as multivariate data analysis (MDA), is commonly utilized in various domains, such as assessing Halal product quality control and determining chemical parameters to evaluate the authenticity

2.2 Chromatography for Halal Authentication

Table 2.2 Analysis of adulterants using different types of GC techniques

Adulterant	Food/beverage	GC type	References
Pork	Meat and sausage	GCMS-HS	Nurjuliana et al. (2011)
Lard	Animal fats	GC x GC-TOF-MS	Indrasti et al. (2010)
Alcohol	Beverages red, white, and cooking wine, industrial alcohol (ethanol) made from sugarcane	GC-FID GC x GC-TOF-MS	Wang et al. (2003) Abdul Hamid et al. (2009)
Low-quality meat mixing	Duck meat in mutton	GC-MS	Wang et al. (2003)
Rat meat fat		GC-MS	Guntarti et al. (2021)

of these items. GC is frequently used to assess product purity or to discriminate between the various components of a mixture (Table 2.2).

It can detect oil and fat-based meals based on their fatty acid content. Fatty acid composition is a reliable sign for identifying adulteration. The C20:2 fatty acid markers are utilized in GC analysis to detect pigs in processed meat products. The detection limit for pork has been reported to be 2% (Sawaya et al. 1990). GC-MS can detect adulteration by analyzing a volatile chemical obtained using solid-phase microextraction (SPME) (Sawaya et al. 1990). SPME and GC can detect volatile chemicals in pork to determine its Halal or kosher status. Estevez et al. studied volatile chemicals in various meats from Iberian and lean pigs, including raw, chilled, cooked, and chilled cooked. The 57 preliminarily discovered volatiles were divided into 11 classes. The volatile compound profiles of both forms of cooked pork were similar, albeit with quantitative differences. Volatile lipids include acids (nonanoic acid), ketones (octane-2,3-dione), aldehydes (pentanal, hexanal, heptanal, octanal, nonanal, decanal, het-(E)-2-enal, dec-(E)-2-enal, deca-(E, Z)-2,4-denial, and undec-(E)-2-enal), furans (2-pentyl-furan), and alcohols (hexane-1-ol, oct-1) (Estévez et al. 2003).

GC and SPME can detect volatile chemicals in cooked, frozen, and raw pork, distinguishing between different types based on enzymatic and oxidative deterioration, fat content, and compositional parameters (Estévez et al. 2003). Using EN and GCMS-HS (Headspace Analyzer), different meats and sausages (beef, mouflon, chicken) can be identified from pig and sausage (Nurjuliana et al. 2011). Preparing samples for GC analysis can be time-consuming and challenging. To identify fatty acid content, methylation processes are required to convert fatty acids to methyl esters (Marikkar et al. 2002). Laboratory equipment and chemicals are needed for several tasks such as mobile phase, drying, and hoover.

Chemometrics offers robust tools for extracting valuable information from large datasets generated by instrumental studies, including spectroscopic and chromatographic approaches. The Chemometrics techniques commonly employed in product authentication can be categorized into many key areas: exploratory data analysis,

Table 2.3 Types of adulterants analyzed using HPLC

Type of food	Adulterant	References
Processed food	Horse and pork meat	von Bargen et al. (2014)
Vegetable oil	Lard	Marikkar et al. (2005)
Fried chicken	Lard	Marikkar et al. (2003)
Gelatins	Bovine gelatin, porcine gelatin, and fish gelatin	Azilawati et al. (2015)
Meat	Beef, mutton, chevon, and chicken	Jorfi et al. (2012)

data pre-processing, description and visualization, discrimination and pattern recognition (classification), regression and prediction, and experimental design. Several challenges encountered during the Halal authentication analysis using chromatography. These challenges included the assessment of the quality of separation, the evaluation of peak alignment through pre-processing techniques, the optimization of chromatographic systems to ensure adequate separation of all peaks using experimental design methods, the accuracy of discrimination and classification using pattern recognition techniques, and the application of multivariate calibration for quantitative analysis. The application of chemometrics in the pre-processing stage has been extensively utilized to achieve the appropriate analytical modeling. HPLC utilizes different types of detectors that are extensively used in food authentication and Halal authentication (Table 2.3).

HPLC with a fluorescence detector can detect markers like hydroxyproline and amino acids in gelatin and collagen samples. This can be used to verify the source of gelatin and collagen. Hydroxyproline is present in different amounts in gelatin and collagen except in fish samples. There are various approaches for Halal authentication, the first being an identification of non-Halal biomarkers, which are distinctive compounds present in non-Halal sources; a standard is required for this approach. The second approach is fingerprinting; in this approach, fingerprint chromatograms of non-Halal sources are compared with the sample. The third approach is metabolomics-based; it utilizes targeted and untargeted techniques to analyze all the metabolites in a sample to be investigated. Chemometrics is mainly used in the second and third approaches, which makes complex and large data more understandable (Nurani et al. 2022).

HPLC detection involves separating the analyte (sample) from the eluent (mobile phase) using a stationary column. The separation is achieved while preserving the sample's molecular structure.

LC analysis is performed for food authentication of meat components for species authentication and adulteration detection (Chou et al. 2007). With the help of different detectors, HPLC can quantify and detect proteins, peptides, and amino acids (von Bargen et al. 2014). HPLC is used to analyze meat to detect histidine dipeptides, balenine, carnosine, and anserine to authenticate species (Aristoy and Toldrá 2004). HPLC-based detection and quantification of triacylglycerol (TAG) is common practice for species identification (Szabo et al. 2007). TAG is used to detect lard in certain food products, i.e., oils and fats (Marikkar et al. 2005). It is essential to note that the chemistry of pork fat is differs in the TAG location. Therefore, HPLC can be used to check the quality of palm, canola, and palm kernel oils.

The emergence of untargeted metabolomics has been recognized as a promising method in metabolomics analysis for food authentication, specifically in Halal examination. Untargeted metabolomics can be used to identify known and unknown components in food. While using untargeted metabolomics, a broad range of metabolites can be identified; this is useful for the authentication of food samples. Non-specific methods can be used to detect specific metabolites in specimens through data analysis. This technique has the advantage of detecting species types, geographic locations of species, and other genetic information in food samples. Untargeted metabolomics is also called a fingerprinting technique used for metabolite identification. This technique is beneficial in identifying non-Halal sources of meat in Halal meat. Sample preparation is rather complex in metabolomics compared to chromatographic techniques. Analysis results can vary widely depending on the nature and condition of the sample.

For targeted metabolomics, extraction of selective metabolites from complex metrics of samples is often required, while in untargeted metabolomics, extraction of a wide variety of metabolites is achieved from the sample. Due to compounds' complex chemistry and charged-based properties, extracting metabolites is often difficult. Certain solvents, i.e., methanol and acetonitrile, are mainly used to extract metabolites in metabolomics. These solvents can be used to extract metabolites of different polarities selectively. Different polarities of these solvents are achieved by utilizing different ratios of methanol acetonitrile in combination with water. Classic techniques such as the Folch method are commonly used for lipid extraction in research. These methods are capable of extracting different types of lipids from samples. Two-phase extraction utilizing multiple solvents is mainly used. The latest advancement in extraction technology has created a phase extraction approach for lipid extraction. This technique is superior to the two-phase extraction technique in cost and efficiency. One-phase extraction techniques for lipids can be achieved with many solvents like methanol, dichloromethane, chloroform, isopropanol, and methyl tertiary butyl ether. Analysts can access and utilize several biomarkers databases to identify non-Halal meat components and other related applications, for example, Comet, X bye Tandem, Proteom Discoverer, and Protein Pilot. Table 2.4 shows the merits and demerits of the LC and GC techniques.

2.3 Mass Spectroscopy

Mass spectroscopy (MS) is an analytical technique in which a sample is converted into ions, and the mass-to-charge ratio of these ions is measured. It is widely employed in several disciplines, i.e., chemistry, biology, environmental science, and food science. MS can identify complex mixtures' components and is considered the best technique for structural determination. This technique can identify and

Table 2.4 Merits and demerits between GC-MS and LC-MS

Characteristic	GC-MS	LC-MS
Technology robustness	Robust and rugged	Super sensitivity
Cost	Relatively inexpensive compared to LC-MS/MS	Expensive instrumentation
Qualitative and quantitative analyses	Yes	Not very quantitative
Sample size	Modest	Minimal
Sensitivity	Good	Very high
Metabolite identification resources	Large database and software availability	Limited database and software resources
Detectable molecules	Most organic, some inorganic	Most organic, some inorganic
Separation reproducibility	Excellent	Poor resolution and reproducibility vs. GC
Sample recoverability	No	No
Sample derivatization	Required	Not required
Separation required	Yes	Can be done without separation (direct injection)
Analysis speed	Slow (20–30 min/sample)	Slow (20–30 min/sample)
Imaging capability	Not applicable	Possible, e.g., MALDI for metabolite imaging
Instrumentation robustness	High	Less robust than GC-MS
Novel compound identification	Difficult	Difficult

Aini et al. (2023)

quantify different types of compounds in different types of metrics. This technique is regarded as the gold standard for determining the structure of compounds and ratios of isotopes in a sample. There are four basic steps in MS Analysis: preparation of the sample, ionization of the sample, mass analysis of the sample, and finally, detection. The sample matrix should be processed in a form suitable for this technique's analysis. Extraction, purification, and concentration of the sample are required to remove interfering compounds and increase the detectability of the analytes. Solid phase or liquid-liquid extraction is mainly employed to remove unwanted compounds; sometimes, derivatization is also performed.

MS utilizes different methods to ionize samples. Electron impact ionization is a method in which high-energy electrons are bombarded on the sample, generating radical cations. This method is helpful for highly volatile and nonpolar compounds. This technique has the drawback of causing fragmentation of large molecules, which can negatively impact the analysis results. Electrospray ionization (ESI) converts the liquid form of samples into charged droplets, and the solvent of these droplets is removed, and charged particles are collected and detected according to their mass-to-charge (m/z) ratio. This method is helpful for the analysis of large biomolecules as it retains the integrity of proteins, peptides, and other biomolecules, and their structure and mass can be detected with this method. However, ESI is not suitable for volatile compounds. The third method is matrix-assisted laser desorption ionization (MALDI). This technique uses metrics in which the sample is present, and a laser is used to ionize the sample in the matrix. This process leads to the desorption and ionization of the sample. This technique helps ionize and analyze large biomolecules to measure their m/z. MALDI cannot be used for smaller compounds. ESI is compatible with LCMS, but MALDI is considered most suitable for tandem MS.

A major component of MS is the mass analyzer, which is responsible for separating and classifying ions according to their mass-to-charge ratio. There are different types of mass analyzers available, and they have various advantages and disadvantages. A Quadrupole mass analyzer consisting of four rods is used to investigate other types of compounds. This mass analyzer applies radio frequency and electric fields between parallel rods to attract charged particles. The electric field in the mass analyzer can selectively filter and classify ions according to their m/z. The second class of mass analyzers is the time of flight (TOF) mass analyzer, based on the principle of how much distance a charged particle can travel depending upon its master charge ratio. This technique is very precise and used for high-resolution separation. TOF analyzers are considered better than quadrupole mass analyzers. TOF mass analyzers can achieve higher resolution and precise measurement of the time of flight of ions. TOF mass analyzers can handle various ions in terms of mass-to-charge ratio, so their application is much wider. Another class of mass analyzers are ion trap mass analyzers, which are capable of trapping ions for longer periods. This feature makes them capable of running multiple scans or analyses of a single ion in more complex samples. Ion trap mass analyzers are smaller and more portable compared to the TOF analyzers. The TOF analyzers can have high throughput applications; however, iron trap analyzers can handle complex samples. Ion trap analyzers, such as linear ion traps or 3D ion traps, come in various types and can precisely measure samples' mass and charge ratio.

2.4 ELISA

Chromatography and electrophoresis are high-end techniques that can authenticate and analyze various food samples. Despite many applications and capabilities, these techniques are expensive, time-consuming, and complex. Recently, genetic and immunological technologies have analyzed the identification and authentication of meat, dietary products, processed food, and cooked food. Genetic methods are considered specific, rapid, and cost-effective analysis methods for food components. Chromatography and electrophoresis require large and expensive instruments and very skillful personnel to operate, while immunological methods are simple and comparatively inexpensive in analyzing and retrieving food. There are various methods of immunological assays, but enzyme-linked immunosorbent assay (ELISA) is the most widely used technique (Table 2.5).

This is due to simplicity, specificity, and cost-effectiveness. There are different types of ELISA, but two types are most common; one is indirect ELISA, and the

Table 2.5 Halal authentication analysis using ELISA

Type of food	Adulterant	References
Raw and heat-processed meat	Pork	Liu et al. (2006)
Raw and heat-processed meat	Beef, pork, poultry, sheep, horse, deer (species authentication)	Ayaz et al. (2006)
Meat	Poultry, rat, kangaroo, and horse (species authentication)	Renčová et al. (2000)

other one is sandwich ELISA. The indirect ELISA utilizes two antibodies that react with the specimen and enzyme. Due to the use of enzymes, this technique is called enzyme-linked. An enzyme is used to create a reaction that causes the generation of a signal, which is detected by chromogenic or fluorogenic substrate. Sometimes, the secondary antibody is linked with proteins such as avidin, and the primary antibody is tagged with biotin.

In a sandwich, ELISA antigen is sandwiched between two antibodies; one is called a capture antibody, and the other is a detection antibody. Detection antibody is linked with enzymes and produce signals. ELISA is capable of qualitative and quantitative analysis; an analyst statistically analyzes the results obtained. ELISA performs qualitative analysis of fluorescent compounds. A standard is produced by serial dilution of the sample, and the results are interpolated to the plot curve. The target analyte should be in sufficient concentration to be detected by antibodies. Both types of antibodies, monoclonal and polyclonal, are used in ELISA. Polyclonal antibodies are considered more specific and can bind different types of epitopes of an antigen. They can detect minorly altered antigens and denatured antigens as well. They are considered the preferred option for detecting denatured proteins. They have limitations like varying affinity, and require a high level of purified sample. Hybridoma technology mainly synthesizes monoclonal antibodies and possesses definite biological activity and consistent specificity.

Polyclonal and monoclonal antibodies (MAbs) are utilized in the several ELISA versions previously documented for food authentication. In addition, it should be noted that the ELISA methodology has a significant drawback in that the target proteins may undergo denaturation during food processing, resulting in the absence of the target protein epitope that the antibodies can detect. However, this limitation has been addressed by creating antibodies specifically designed to recognize and bind to thermo-stable proteins (Asensio et al. 2008).

2.5 DNA-Based Techniques

DNA-based technologies offer numerous possibilities for researchers, managers, and regulators engaged in identifying, quantifying, and surveilling contaminated species in meat and meat products. The surge in attention toward advancing DNA-based techniques for species identification has sparked extensive conjecture

2.5 DNA-Based Techniques

regarding the potential future accessibility of cost-effective, swift, and precise methods for identifying and quantifying all declared or undisclosed constituents in finalized commercial goods. DNA-based techniques have demonstrated their efficacy in detecting even trace amounts of adulteration in extensively processed beef products, offering promising potential for enhancing transparency and promoting fair-trade practices within the food business. Using polymerase chain reaction (PCR) has facilitated the retrieval and amplification of DNA from specific species, enabling the detection of even small quantities of DNA in heavily degraded materials or a complex mixture of genetic material. The presence of consistent information content enhances the method's sensitivity. Multiple copies of specific nucleic acids, such as mitochondrial DNA (mtDNA), are ubiquitously present, and polymorphic features are more extensively distributed across the genome.

The techniques above can identify the species inside intricate and extensively processed food matrices, even when the DNA is fragmented and hence unsuitable for accurate identification. In order to determine the DNA of various species in heavily processed food, the PCR methods have been enhanced by focusing on DNA fragments of shorter length (less than 150 base pairs). These shorter fragments exhibit high stability even when subjected to rigorous food processing procedures. The selection of a particular gene is undertaken to construct a primer set specific to the species being studied, ensuring the PCR assay's success. The choice of gene is determined by the presence of conserved regions across different species and variable regions within the same species. These regions contribute to enhancing the specificity of the PCR experiment. Several PCR approaches have been developed and validated to accurately detect and distinguish various species in raw and processed food matrices, including those undergoing extensive processing (Table 2.6).

Food authentication employs two primary categories of PCR techniques: endpoint (traditional) PCR and real-time PCR. Both methods can accommodate simplex and multiplex systems. The simplex system is a detection approach focusing on identifying a single species, whereas the multiplex system can identify many species inside a single test platform.

Table 2.6 Halal authentication using different types of PCR

Type of food	Analysis	References
Meat	Cat, dog, and rat or mouse (species identification)	Martín et al. (2007)
Meat	Pig, cattle, goat, buffalo, and sheep	Kumar et al. (2014)
Meat	Pork, goat, beef, buffalo, chicken, rabbit, and quail	Murugaiah et al. (2009)
Meat	Dog, cat, horse, and donkey	Abdel-Rahman et al. (2009)
Cooked and processed food	Chicken, beef, mutton, and pork	Zhang (2013)
Boiled, autoclaved, and microwave cooked food	Cow, buffalo, chicken, cat, dog, pig, and fish	Hossain et al. (2019)

2.6 Electronic Nose

The evaluation of the quality of many products in everyday life, such as foods and cosmetics, heavily relies on the human nose's sense of volatile chemicals. Hence, it is unsurprising that numerous attempts have been made over time to develop instruments that function based on a comparable concept to that of the human nose. In most instances, these systems would not serve as a substitute for, but rather complement, conventional assessments of volatile chemicals conducted by sensory methods and standard analytical procedures (Table 2.7).

The proposition of an artificial olfactory system was introduced in 1982 at the University of Warwick by Persaud and Dodd. In essence, these systems necessitate the utilization of gas sensors, which were initially devised over three decades ago. During the early 1990s, the terminology of "artificial" or "electronic nose" emerged, introducing many commercially accessible instruments. Further extensive study was initiated, allowing for the exploration and evaluation of applications, particularly within the food business.

According to Gardner and Bartlett, the electronic nose can be characterized as an instrument consisting of an array of electronic chemical sensors with limited specificity and a suitable pattern-recognition system. This device is capable of identifying both basic and complex scents. The analytical system under consideration significantly differs from the human olfactory system, as it is characterized by its electronic nature rather than a true olfactory organ. The sole similarity between our olfactory organ and the aforementioned entity lies in their shared functionality. Similar to the olfactory system seen in mammals, this sensory apparatus can detect various gases through specialized sensors. These sensors transmit signals to a recognition organ, such as the brain or a computer, for further processing and analysis. However, the operating principle, the quantity of sensors, and the levels of sensitivity and selectivity vary significantly. This is the reason why certain scientists opt to refer to this device using alternative terms such as "flavor sensor," "aroma sensor," "odor-sensing system," or "multi-sensor array technology."

Electronic nose systems consist of a hardware component that is typically equipped with:

I. Sensors
II. Electronics
III. Pumps

Table 2.7 Applications of electronic nose in Halal food analysis

Type of food	Analysis	References
Pork, chicken, mutton, beef, pork sausage, chicken sausage, and beef sausage	Food authentication	Nurjuliana et al. (2011)
Ethanol content in soy sauce	Quantitative and qualitative analysis	Park et al. (2017)
Minced pork and minced mutton	Pork adulteration detection	Sarno et al. (2020)

IV. Air conditioning
V. Flow controllers

Additionally, software is designed to monitor the hardware, pre-process data, perform statistical analysis, and perform other related tasks. In order to function effectively as analytical instruments, these systems must be designed to be utilized over extended periods while maintaining a high level of repeatability, which refers to the ability to consistently obtain the same pattern for a sample on the same array within short time intervals. Additionally, these systems should also exhibit reproducibility, meaning that different sensor batches or instruments should be capable of producing identical patterns for the same sample. Sensor technology plays a pivotal role in achieving the desired analytical properties. Specific mention is then given to using electronic nose systems in food-related applications (Schaller et al. 1998).

2.7 Sensors

Sensors of various types demonstrate interactions with the gas under measurement, resulting in a sequence of physical and/or chemical interactions as volatile chemicals pass over the sensor. A state of dynamic equilibrium arises when volatile compounds undergo continuous adsorption and desorption processes at the sensor's surface.

The sensors that are most suitable for integration in an electronic nose should meet the following criteria: They should exhibit a high level of sensitivity towards chemical compounds comparable to that of the human nose (reaching a detection limit of 10–12 g/mL); they should have low sensitivity towards humidity and temperature; they should possess a moderate level of selectivity, enabling them to respond to various compounds present in the sample's headspace; they should demonstrate high stability, reproducibility, and reliability; they should have short reaction and recovery times; they should be robust and durable; they should be easily calibrated; they should provide easily processable data output; and they should have small dimensions.

The sensors are specifically engineered for industrial applications, particularly in online systems. They offer several valuable advantages, including minimal operational temperature, low power consumption, high safety, and cost-effective production. Most firms are actively seeking sensors that exhibit a high degree of selectivity. In the context of an electronic olfactory system, each sensor must be capable of detecting any molecule in the gaseous phase. If a novel component is introduced into a mixture, at least one sensor must be capable of detecting this specific addition. The excessive deployment of sensors results in an overly intricate system characterized by an abundance of superfluous data.

Multiple gas sensors exist; however, only four technologies are presently employed in commercialized electronic noses. These technologies include metal oxide semiconductors (MOS), metal oxide semiconductor field effect transistors

(MOSFET), conducting organic polymers (CP), and piezoelectric crystals (bulk acoustic wave = BAW).

Fiberoptic, electrochemical, and bi-metal sensors, among others, are now undergoing development and hold potential for integration in future iterations of electronic noses.

The sensors can be categorized into two primary classes: hot sensors (MOS and MOSFET) and cold sensors (CP, SAW, and BAW). The former devices function at elevated temperatures and are regarded as being less susceptible to moisture, resulting in reduced carry-over between successive measurements. They provide balanced sensitivity between drift and longevity.

Metal oxide-based semiconductors are gas and volatile organic compounds sensors. It was first introduced in Japan by Taguchi for application in gas-sensing alarms. Their sensing is based upon changes in the conductivity of metal oxide when they react with gases or volatile organic compounds.

The construction of metal oxide-based sensors is very simple. They are composed of base material, ceramic and metal oxide, wrapped around these ceramic materials, and this setting is connected to a heating wire. The metal oxide coating is of two types: one is p-type, and another is n-type. N-type comprises zinc oxide, tin dioxide, or titanium dioxide, and p-type is composed of nickel or cobalt oxide.

Photolytic excitation in a n-type semiconductor (where n represents negative electrons) or donor nature leads to an accumulation of additional electrons in its conduction band. This surplus of electrons enhances the semiconductor's reactivity with oxidizing molecules. A p-type semiconductor, also known as a positive hole or acceptor semiconductor, exhibits an enthusiastic state characterized by an electron deficit in its valence band. This deficit facilitates interactions with reducing chemicals.

The film deposition process categorizes each type of sensor into thin (6–1000 nm) or thick (10–300 μm) film MOS sensors. Film deposition involves several techniques to produce thin films, such as physical or chemical vapor deposition, evaporation, and spraying. Conversely, methods such as screen printing or painting are employed to create thick films. Thin film devices provide enhanced response times and substantially increased sensitivities but at the cost of heightened manufacturing complexities about reproducibility. Hence, MOS sensors are readily accessible in the market and rely on thick film technology.

The resistance of the metal oxide is altered due to the complete combustion of organic volatiles on its surface, which occurs due to the elevated operating temperature range of 200–650 °C. This combustion process converts the organic volatiles into carbon dioxide and water. The underlying mechanism of the two-step reaction involves an oxygen exchange process between the volatile molecules and the metal layer. Initially, the oxygen molecules in the carrier gas are adsorbed onto the material's surface. Subsequently, these molecules penetrate the metal coating and occupy the voids within the polycrystalline lattice structure of the metal. The electrons experience an attractive force exerted by the oxygen that has been charged, resulting in a drop in sensor conductivity. Furthermore, gas and surface interactions occur through two mechanisms: the adsorbed volatile molecules and the metal coating. (i)

At lower temperatures, charges are transferred between the volatile substances and the oxygen that is adsorbed. (ii) At higher temperatures, the oxygen within the lattice reacts with the adsorbed volatile compounds, resulting in a change in the concentration of defects within the crystal. The differentiation between both pathways is not readily apparent, and both can occur concurrently depending on whether the electrophilic characteristics of chemisorbed oxygen (O_2 ads)— and (O ads)—, or the nucleophilic characteristics of lattice oxygen O_2—are implicated.

In order to modify the preferential reactivity of a metal oxide film towards various chemical compounds, the film is subjected to doping with noble catalytic metals such as platinum or palladium; alternatively, the operating temperature is adjusted within the range of 50–400 °C. The particle size of the polycrystalline semiconductor affects the selectivity of MOS sensors. However, they often exhibit lower selectivity than other technologies such as CP, BAW, SAW, or MOSFET. MOS sensors exhibit a high sensitivity towards ethanol, resulting in their inability to detect other volatile compounds of interest. At higher temperatures, these materials tend to be unsuitable for combustible materials.

Organic polymers have shown substantial growth in the last decade. These sensors utilize changes in resistance to detect gaseous materials. The substrate of these sensors is made up of silicon, glass, or fiberglass, which is connected to a pair of electrodes. The electrodes are gold coated; a sensing material made up of organic polymers (polypyrrole, polyaniline, or polythiophene) is also added. The polymer layer is deposited using electrochemical desorption. Polymers contain cationic sites, while anions are present in electrolytes or solvents.

The cation sites are likely composed of polarons or bipolarons, localized pockets of positive charge within the polymer chain. These regions facilitate the movement of electrons, hence enabling electron transport. An electric current flows through the conducting polymer when an electrical potential difference is applied across the electrodes. Introducing volatile substances onto the sensor surface induces changes in the electron flow within the system, consequently affecting the sensor's resistance. The volatiles have the potential to interact with several components, including (i) the polymer itself, (ii) the counterion, or (iii) the solvent (Tao et al. 2018). Hence, the attainment of desirable selectivity in CP sensors can be accomplished by modifying either one of these factors or the electrical development of the polymer coating.

Overall, these sensors exhibit favorable sensitivities, particularly concerning polar chemicals. Nevertheless, these devices' low operating temperature (below 50 °C) renders them highly susceptible to moisture. Despite their resistance to poisoning, these sensors have a limited lifespan of approximately 9–18 months. The limited lifespan of the sensor may be attributed to the process of polymer oxidation or the potential impact of various substances on the development of contact resistances between the polymer and the electrodes.

In contrast to MOS sensors, CP sensors have not yet achieved widespread market availability, mostly due to the high costs associated with laboratory-scale fabrication. The primary drawbacks of these systems are the challenges related to achieving consistent batch-to-batch repeatability and the tendency for the response to exhibit drift over time.

Piezoelectric crystal sensors utilize the piezoelectric effect to convert mechanical energy into electrical signals. These sensors are commonly used in various applications. For example, piezoelectric sensors operate by detecting alterations in mass, which can be quantified as variations in resonance frequency.

The sensors in this context consist of minute discs, often composed of quartz, lithium niobate ($LiNbO_3$), or lithium tantalate ($LiTaO_3$). These discs are coated with chromatographic stationary phases, lipids, or any chemically and thermally stable non-volatile chemicals.

When an alternating electrical current is applied at ambient temperature, the crystal exhibits a highly consistent vibration frequency determined by its mechanical characteristics. When the coating is exposed to a vapor, it adsorbs specific molecules, increasing the sensing layer's mass. As a result, the resonance frequency of the crystal drops. This potential alteration could be subject to surveillance and connected to unstable circumstances.

The crystals can generate vibrations in a bulk acoustic wave (BAW) or surface acoustic wave (SAW) mode. This can be achieved by carefully choosing the optimal crystal cut and electrode design (Länge 2019).

There are structural differences between BAW and SAW sensors. The phenomenon known as BAW refers to three-dimensional waves that propagate within a crystal. On the other hand, SAW encompass two-dimensional waves, specifically Rayleigh, Love, and Bluestein-Gulyaev waves, which propagate over the crystal's surface at a depth roughly equivalent to one wavelength. The Piezoelectric Sorption Detector, developed by King in 1964, introduced the notion of BAW sensors.

These devices are commonly referred to as "quartz crystal micro-balance" (QCM or QMB) due to their resemblance to a balance, as their responses vary concerning the quantity of mass adsorbed. BAW sensors exhibit vibrations at a frequency range of 10–30 MHz. The application of a thin layer (ranging from 1 μm to 10 nm) is achieved by techniques such as spin coating, airbrushing, or inkjet printing.

In 1979, Wohltjen and Dessy introduced the first gas sensor with a SAW oscillator. This development occurred over a decade after the initial timeframe. The utilization of said sensor was initially documented by Martin et al. during the 1980s.

These devices function at a higher frequency range than BAW sensors, specifically from 100 MHz to 1 GHz. The manufacturing approach employed in this process involves the utilization of photolithography and airbrushing. It is worth noting that this technique is entirely compatible with creating planar integrated circuits, particularly those utilizing planar silicon technologies. Integrating SAW structures and conditioning circuits on a single silicon substrate allows for the development of durable and cost-effective SAW sensors.

Piezoelectric sensors offer superior selectivity because they can be coated with a wide range of materials. Nevertheless, the current state of coating technology lacks adequate control, resulting in subpar consistency between batches. Although constrained by the noise generated due to their elevated operational frequency, saw sensors have greater sensitivity than BAW sensors. Nevertheless, it should be noted that both sensors exhibit a greater sensitivity to volatiles, necessitating a higher concentration of such volatiles to achieve response levels comparable to those observed in other types of sensors.

The challenge associated with integrating BAW and SAW sensors into an electronic olfactory system lies in the intricate electronic components and their susceptibility to external factors, such as variations in temperature and humidity.

MOSFET sensors are electronic devices that utilize the principles of MOSFET technology for sensing applications.

The PdMOS (palladium metal oxide semiconductor) device, which exhibits sensitivity to hydrogen, was initially invented in 1973 by Swedish researchers. The findings of this study were then published 2 years later by Lundström et al. The sensors based on MOSFETs operate by detecting alterations in electrostatic potential. The MOSFET sensor is composed of three distinct layers: a silicon semiconductor, a silicon oxide insulator, and a catalytic metal (often palladium, platinum, iridium, or rhodium), which is commonly referred to as the gate. A conventional transistor functions through three contacts, wherein two of these contacts facilitate the inflow (source) and outflow (drain) of current, while the third contact serves as the gate contact responsible for regulating the current passing through the transistor. The presence of a diode mode characterizes the MOSFET transistor due to the direct connection between the gate and drain contacts. This feature provides convenient electronic operation and can be represented by an IV-curve.

Applying power to the gate and drain contacts generates an electric field that impacts the conductivity of the transistor. Modification of the electric field and the resulting alteration of current flowing through the sensor occur when polar chemicals interact with the metal gate.

The recorded response indicates the voltage adjustment required to maintain a consistent predetermined drain current. The gate structure of a MOSFET sensor can be categorized into two types: a thick, dense metal film with a thickness ranging from 100 to 200 nm or a thin, porous metal film with a thickness ranging from 6 to 20 nm. The metal gate, which is dense and uninterrupted, primarily reacts to molecules that undergo hydrogen dissociation on the catalytic metal surface. The assumption is that the insulator is not directly exposed to the surrounding molecules. The hydrogen atoms that have undergone dissociation rapidly disperse within microseconds throughout the metal, forming a dipole layer at the interface between the metal and insulator. This dipole layer induces a change in potential within the transistor. Detecting compounds such as ammonia or carbon monoxide is unattainable with this particular layer due to the absence of hydrogen atom release.

However, it was discovered that the latter compounds exhibit favorable responses when the metal gate is reduced in thickness. The most likely mechanism can be attributed to three voltage shifts: (i) occurring at the metal surface as a result of adsorption and chemical reactions, (ii) taking place at the metal-insulator interface due to hydrogen diffusion through the metal, and (iii) happening at the insulator surface due to reactions involving polar molecules or charges on the oxide metal surface, which either directly react with the oxide metal or diffuse out on the surface.

The selectivity and sensitivity of MOSFET sensors can be affected by various factors, including the operating temperature within the range of 50–200 °C, the metal gate's composition, and the catalytic metal's microstructure. MOSFET sensors, similar to MOS sensors, exhibit a comparatively diminished responsiveness to moisture and are often regarded for their high durability. However, significant manufacturing expertise is necessary to achieve good quality and reproducibility.

References

Abdel-Rahman S et al (2009) Detection of adulteration and identification of cat's, dog's, donkey's and horse's meat using species-specific PCR and PCR-RFLP techniques. Aust J Basic Appl Sci 3(3):1716–1719

Abdul Hamid N et al (2009) Identification of alcoholic compounds in fermented glutinous rice (Tapai). In: 3rd IMT-GT international symposium on Halal science and management

Aini S et al (2023) The metabolomics approach used for Halal authentication analysis of food and pharmaceutical products: a review. Food Res 7(3):180–187

Aristoy MC, Toldrá F (2004) Histidine dipeptides HPLC-based test for the detection of mammalian origin proteins in feeds for ruminants. Meat Sci 67(2):211–217

Asensio L et al (2008) Determination of food authenticity by enzyme-linked immunosorbent assay (ELISA). Food Control 19(1):1–8

Ayaz Y et al (2006) Detection of species in meat and meat products using enzyme-linked immunosorbent assay. J Muscle Foods 17(2):214–220

Azilawati M et al (2015) RP-HPLC method using 6-aminoquinolyl-N-hydroxysuccinimidyl carbamate incorporated with normalization technique in principal component analysis to differentiate the bovine, porcine and fish gelatins. Food Chem 172:368–376

Chou C-C et al (2007) Fast differentiation of meats from fifteen animal species by liquid chromatography with electrochemical detection using copper nanoparticle plated electrodes. J Chromatogr B 846(1–2):230–239

Estévez M et al (2003) Analysis of volatiles in meat from Iberian pigs and lean pigs after refrigeration and cooking by using SPME-GC-MS. J Agric Food Chem 51(11):3429–3435

Guntarti A et al (2021) Authentication of Sprague Dawley rats (Rattus norvegicus) fat with GC-MS (gas chromatography-mass spectrometry) combined with chemometrics. Int J Appl Pharm 13(2):1–6

Hossain MM et al (2019) Heptaplex polymerase chain reaction assay for the simultaneous detection of beef, buffalo, chicken, cat, dog, pork, and fish in raw and heat-treated food products. J Agric Food Chem 67(29):8268–8278

Indrasti D et al (2010) Lard detection based on fatty acids profile using comprehensive gas chromatography hyphenated with time-of-flight mass spectrometry. Food Chem 122(4):1273–1277

Jaswir I et al (2017) An overview of the current analytical methods for Halal testing. In: Contemporary issues and development in the global Halal industry: selected papers from the international Halal conference 2014. Springer

Jorfi R et al (2012) Differentiation of pork from beef, chicken, mutton and chevon according to their primary amino acids content for Halal authentication. Afr J Biotechnol 11(32):8160–8166

Kisley L, Landes CF (2015) Molecular approaches to chromatography using single molecule spectroscopy. Anal Chem 87(1):83–98

Kumar D et al (2014) Authentication of beef, carabeef, chevon, mutton and pork by a PCR-RFLP assay of mitochondrial cyt b gene. J Food Sci Technol 51:3458–3463

Länge K (2019) Bulk and surface acoustic wave sensor arrays for multi-analyte detection: a review. Sensors 19(24):5382. https://doi.org/10.3390/s19245382

Liu L et al (2006) Sensitive monoclonal antibody-based sandwich ELISA for the detection of porcine skeletal muscle in meat and feed products. J Food Sci 71(1):M1–M6

Marikkar J et al (2002) The use of cooling and heating thermograms for monitoring of tallow, lard and chicken fat adulterations in canola oil. Food Res Int 35(10):1007–1014

Marikkar J et al (2003) Lard uptake and its detection in selected food products deep-fried in lard. Food Res Int 36(9–10):1047–1060

Marikkar J et al (2005) Distinguishing lard from other animal fats in admixtures of some vegetable oils using liquid chromatographic data coupled with multivariate data analysis. Food Chem 91(1):5–14

Martín I et al (2007) Detection of cat, dog, and rat or mouse tissues in food and animal feed using species-specific polymerase chain reaction. J Anim Sci 85(10):2734–2739

Murugaiah C et al (2009) Meat species identification and Halal authentication analysis using mitochondrial DNA. Meat Sci 83(1):57–61

Nakyinsige K et al (2012) Halal authenticity issues in meat and meat products. Meat Sci 91(3):207–214

Ng PC et al (2022) Recent advances in Halal food authentication: challenges and strategies. J Food Sci 87(1):8–35

Nurani LH et al (2022) Use of chromatographic-based techniques and chemometrics for Halal authentication of food products: a review. Int J Food Prop 25(1):1399-1416 %@ 1094-2912

Nurjuliana M et al (2011) Rapid identification of pork for Halal authentication using the electronic nose and gas chromatography mass spectrometer with headspace analyzer. Meat Sci 88(4):638–644

Park SW et al (2017) Analysis of ethanol in soy sauce using electronic nose for Halal food certification. Food Sci Biotechnol 26:311–317

Premanandh J, Bin Salem S (2017) Progress and challenges associated with Halal authentication of consumer packaged goods. J Sci Food Agric 97(14):4672–4678

Renčová E et al (2000) Identification by ELISA of poultry, horse, kangaroo, and rat muscle specific proteins in heat-processed products. Vet Med 45(12):353–356

Sarno R et al (2020) Detecting pork adulteration in beef for Halal authentication using an optimized electronic nose system. IEEE Access 8:221700–221711

Sawaya WN et al (1990) Detection of pork in processed meat: experimental comparison of methodology. Food Chem 37(3):201–219

Schaller E et al (1998) 'Electronic noses' and their application to food. LWT Food Sci Technol 31(4):305–316

Szabo A et al (2007) Fatty acid regiodistribution analysis of divergent animal triacylglycerol samples–a possible approach for species differentiation. J Food Lipids 14(1):62–77

Tao Y, Xu J, Liang ZF, Xiong L, Yang HC (2018) Domain correction based on kernel transformation for drift compensation in the E-nose system. Sensors 18:3209

von Bargen C et al (2014) Meat authentication: a new HPLC–MS/MS based method for the fast and sensitive detection of horse and pork in highly processed food. J Agric Food Chem 62(39):9428–9435

Wang M-L et al (2003) A rapid method for determination of ethanol in alcoholic beverages using capillary gas chromatography. J Food Drug Anal 11(2):3

Wilson JA, Liu J (2011) The challenges of Islamic branding: navigating emotions and Halal. J Islam Mark 2(1):28–42

Zhang C (2013) Semi-nested multiplex PCR enhanced method sensitivity of species detection in further-processed meats. Food Control 31(2):326–330

Chapter 3
Halal Food Authentication Using Chemometrics

3.1 Chemometrics

Chemometrics is the statistical and mathematical computation of the chemical data measurements. The measurement ranged from calculating a Fourier transform interpolation of a spectrum to detecting the pH of hydrogen ion activity. According to the International Chemometrics Society, chemometrics is the science of using mathematical or statistical techniques to relate measurements taken on a chemical system or process to the system's state. Chemometrics has its roots in analytical chemistry, although it is entirely interdisciplinary and has been utilized in many different fields such as environmental chemistry, engineering, metabolomics, pharmaceutical studies, forensics, cultural studies, and food sciences (assessment of adulteration, geographical origin, etc.).

3.1.1 History of Chemometrics

Chemometrics originated in the mid-1960s, evolving alongside advancements in scientific computing and laboratory technology. In 1971, the Swedish word "kemometri" was used for the first time by a combination of the two words chemistry and metri. Prof. Svante Wold first introduced the English term chemometrics (chemo+metrics) and formally published it in 1972 (Swarbrick and Westad 2016). Wold and Kowalski founded the International Chemometrics Society in 1974, solidifying its identity as a distinct field. The field gained momentum in the 1980s with international meetings, dedicated journals, books, and software. Chemometrics is now considered essential in analytical chemistry, with its education widely advocated for modern chemists.

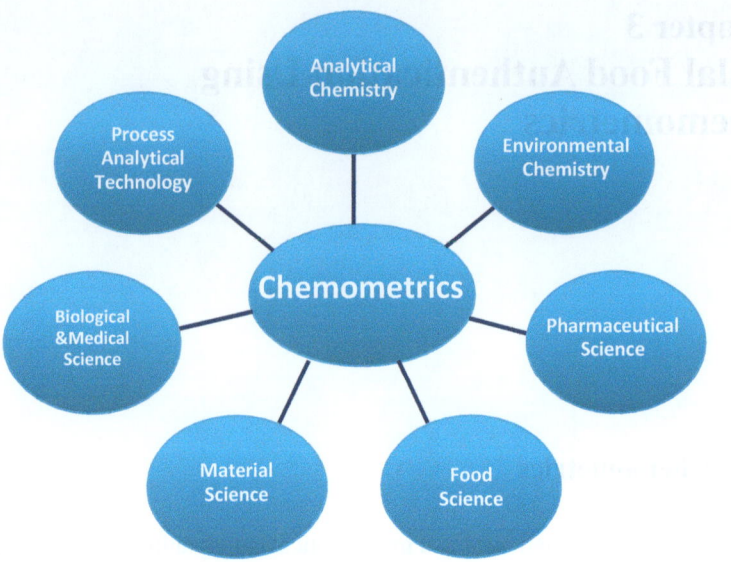

Fig. 3.1 Applications of chemometrics in the scientific field

3.1.2 Chemometrics in Scientific Research

Chemometrics in scientific research is a vital approach that integrates chemistry, mathematics, and computer science to enhance the understanding of complex systems and processes. By analyzing and interpreting data from experiments or measurements, chemometrics helps advance scientific discoveries and optimize methodologies. Figure 3.1 enlists some of the fields where chemometrics is utilized widely.

3.2 Chemometrics in Food Authentication

Chemometrics, particularly multivariate analysis, has numerous uses in quality control and the qualitative and quantitative assessment of chemical parameters for evaluating the authenticity of food products. Chemometrics offers robust methods for the calibration analysis of chromatographic and spectroscopic data, utilized in both targeted and non-targeted methodologies to detect instances of food fraud or to verify their geographic or biological provenance.

In chemometric analysis, there will be a set of samples that need to be characterized in terms of some features that have been measured from them. The measurements can be carried out both on a set of samples of the same origin but with significant natural variations that are normally found in originating sources or on samples of different origins. Generally, the digitized signal spectra of the measurements will contain both spectral information and various degrees of fluctuations in the form of random noise. In these measurements, the features can then be extracted using statistical analysis methods, which involve the use of both pattern recognition and classification of the multivariate relationship between different signals. The development of analytical techniques in chemometrics is mostly due to the advancement of information technology that allows the fabrication of high-level sensor spectral data and the development of a computer that can handle large data matrix calculations (Wang et al. 2022; Liu et al. 2023; Otto 2023; Aleixandre-Tudó et al. 2022; Kharbach et al. 2023; Zhang and Yang 2024; Chapman et al. 2020; Liu and Kazarian 2022; Tortorella and Cinti 2021).

The combination of conventionally existing food authenticity along with modern food science technologies has introduced development in the field of food authentication. In recent years, several authentication detection methods and techniques have been developed for a variety of foods, commonly for origin labeling and the use of genetically modified components in food. Adulteration detection in food is common and has resulted in the development of certain techniques for examining foreign substances in large quantities of food. The traditional method employed by forensic agencies, closely associated with the judicial field, is the analysis involving the physical, chemical, and biological traits of food. Meanwhile, instrumental analysis techniques for food, particularly for food process control, have been employed to detect adulteration. These food control techniques have been designed for specific components and may not be able to detect various food adulterations or the various origins of food (Rohman et al. 2020; Liu et al. 2021; Ferone et al. 2020; Hassoun et al. 2023; Bhat et al. 2022; Zhang et al. 2023; Shen et al. 2021; Rizou et al. 2020).

3.2.1 Univariate Analysis

As the demand for Halal food authentication continues to grow, ongoing research and development in univariate analysis techniques will be essential. Combining these methods with complementary approaches and leveraging emerging technologies will be key to addressing current challenges and enhancing the reliability of Halal food authentication processes. The future of Halal food authentication likely lies in a multifaceted approach, where univariate analysis serves as a critical component within a broader analytical framework. This integrated strategy will be crucial in maintaining consumer trust, ensuring regulatory compliance, and supporting the continued growth of the global Halal food market. Univariate analysis, a statistical method examining one variable at a time, has emerged as a valuable tool in this field. Univariate analysis in Halal food authentication typically involves examining a single attribute or component of a food sample. This can include various analytical techniques, each focusing on a specific marker or characteristic.

Univariate analysis plays a crucial role in Halal authentication of meat, oils, fats, and fish products. Its simplicity, specificity, and often high sensitivity make it a valuable tool in ensuring food integrity and compliance with Halal standards. However, the limitations of univariate approaches necessitate careful method selection, validation, and interpretation of results.

In meat authentication, univariate analysis often focuses on specific biomarkers. For instance, Bargen et al. (2014) used High-Performance Liquid Chromatography (HPLC) to analyze porcine DNA in meat products. This method allows for the detection of pork adulteration in Halal-labeled meats by examining a single DNA marker. Grundy et al. (2016) reported that HPLC analysis could detect porcine DNA at concentrations as low as 0.1% in meat mixtures. This high sensitivity allows for the identification of even trace amounts of pork in Halal-labeled products. Similarly, Rohman et al. (2022) employed Real-Time Polymerase Chain Reaction (RT-PCR) to detect porcine mitochondrial DNA in meat samples. This univariate approach provides a sensitive and specific method for identifying pork contamination in Halal meat products.

For oils and fats, univariate analysis often centers on fatty acid profiles or specific chemical markers. Rohman and Che Man (2011) utilized FTIR spectroscopy to detect lard adulteration in palm oil at levels as low as 1%. This method focuses on the unique spectral patterns of lard, allowing for its identification in oil mixtures and demonstrating the technique's effectiveness in identifying non-Halal components in vegetable oils. In another study, Heidari et al. (2020) used GC-MS to analyze the fatty acid composition of vegetable oils, enabling the detection of non-Halal animal fats in supposedly pure vegetable oils.

Fish authentication often involves species identification to ensure compliance with Halal standards. DNA barcoding a univariate technique focusing on a specific gene region to identify fish species in processed products, has been employed. This method helps prevent the substitution of Halal fish species with non-Halal alternatives. DNA barcoding has shown high accuracy in species identification and fish authentication. Shehata et al. (2018) reported a 95% success rate in identifying fish species in processed products using this univariate approach. Univariate analysis has yielded significant results in Halal food authentication across various food categories. Despite its effectiveness, univariate analysis in Halal food authentication faces several limitations, as shown in Fig. 3.2 and elaborated in the text.

Limited Scope By focusing on a single variable, univariate analysis may miss complex interactions between multiple components. This can be problematic in cases of sophisticated food adulteration where multiple non-Halal ingredients are used (Fadzlillah et al. 2013).

Sensitivity to Processing Some univariate techniques, particularly those based on DNA analysis, can be affected by food processing methods. Heat treatment, for example, can degrade DNA, potentially leading to false negative results (Ali et al. 2015).

Variation in Marker Levels Natural biological variation in the concentration of specific markers can sometimes lead to inconclusive results. This is particularly challenging in plant-based products where environmental factors can significantly influence chemical compositions (Marikkar et al. 2016).

3.2 Chemometrics in Food Authentication

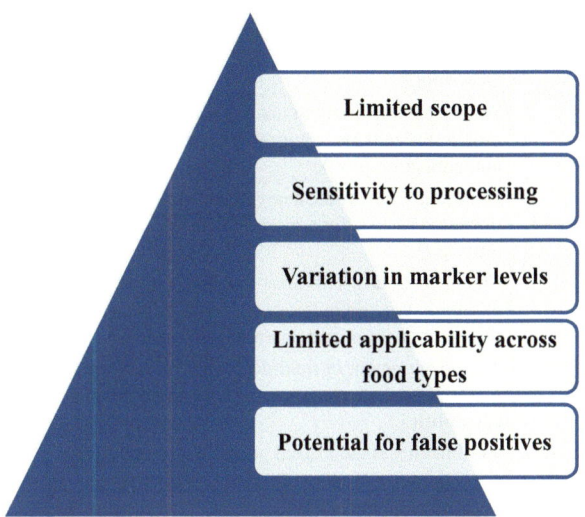

Fig. 3.2 Limitations of univariate analysis in Halal food authentication

Limited Applicability Across Food Types A univariate method developed for one food type may not be directly applicable to others. For instance, a technique optimized for meat authentication may not work effectively for oils or fish (Karahalil 2020).

Potential for False Positives In some cases, univariate analysis may detect compounds similar to the target marker, leading to false positive results. This underscores the need for careful method validation and complementary analytical approaches (Rohman et al. 2011). To address these limitations and further improve univariate analysis in Halal food authentication, several directions are being explored, as sown in Table 3.1.

3.2.2 Multivariate Analysis

Multivariate analysis, which simultaneously considers multiple variables, has emerged as a powerful tool in the authentication of Halal food products, particularly in complex matrices such as meat, oils, fat, and fish. Multivariate analysis in Halal food authentication typically involves the simultaneous examination of multiple attributes or components of a food sample. This approach often employs chemometric techniques to analyze complex datasets generated by various analytical methods.

In meat authentication, multivariate analysis is frequently applied to spectroscopic data. Rohman et al. (2011) used FTIR spectroscopy coupled with Partial Least Squares (PLS) regression to detect and quantify pork adulteration in beef meatballs. This method analyzes the entire spectral profile rather than focusing on a

Table 3.1 Advances in univariate analysis for Halal authentication

Advances in chemometrics	Description	References
Multi-marker approaches	While still univariate in nature, analyzing multiple individual markers can provide a more comprehensive authentication profile	Scarano and Rao (2014)
Integration with multivariate methods	Combining univariate analysis with multivariate techniques can offer a more robust authentication process	Todorov et al. (2020)
Non-targeted screening	Emerging untargeted metabolomics approaches, although more complex, can provide a broader view of food composition and potential adulterants	Cubero-Leon et al. (2014)
Portable and rapid testing	The development of field-deployable univariate testing methods can enhance on-site Halal authentication capabilities	Fathima et al. (2024)
Standardization efforts	Establishing standardized protocols for univariate analysis in Halal authentication can improve consistency and reliability across different laboratories and regions	Hafis Yuswan et al. (2020)

single marker. Similarly, Guntarti et al. (2017) employed GC-MS with Principal Component Analysis (PCA) to differentiate between beef and pork based on their fatty acid profiles. This multivariate approach allows for a more comprehensive comparison of meat samples.

For oils and fats, multivariate analysis often involves the examination of multiple chemical parameters. Che Man et al. (2011) utilized FTIR spectroscopy combined with PCA and PLS to detect lard adulteration in vegetable oils. This method considers the entire spectral fingerprint of the oils, providing a more robust authentication process. In another study, Marikkar et al. (2005) used HPLC data with multivariate analysis to detect the presence of lard in vegetable oils and animal fats. The combination of chromatographic data and chemometric analysis enhanced the ability to detect subtle differences in oil compositions.

Fish authentication often involves the analysis of multiple genetic or protein markers. Mottola et al. (2022) employed DNA metabarcoding coupled with multivariate analysis to differentiate between fish species in processed products. This approach considers multiple genetic loci simultaneously, improving the accuracy of species identification. Martinez and Jakobsen Friis (2004) used proteomic analysis with multivariate statistical tools to authenticate fish species in commercial products. By examining multiple protein markers, this method provides a more comprehensive species identification approach.

Multivariate analysis has demonstrated significant success in Halal food authentication across various food categories. In meat authentication, the FTIR-PLS method developed by Rohman et al. (2011) could detect pork adulteration in beef meatballs with a detection limit of 5%. This high sensitivity allows for the identification of even small amounts of pork in processed meat products.

For oils and fats, Che Man et al. (2011) reported that their FTIR-PCA-PLS method could detect lard adulteration in vegetable oils at levels as low as 1%. This demonstrates the technique's effectiveness in identifying non-Halal components in complex oil mixtures.

In fish authentication, the DNA barcoding with multivariate analysis approach used by Mottola et al. (2022) achieved a 98% success rate in identifying fish species in processed products. This high accuracy highlights the power of combining genetic analysis with multivariate statistical tools. Despite its effectiveness, multivariate analysis in Halal food authentication faces several limitations, as shown in Fig. 3.3 and elaborated in the text one by one.

Complexity of Data Interpretation The multidimensional nature of the data can make interpretation challenging, requiring expertise in both analytical chemistry and statistical analysis (Georgiou and Danezis 2015).

Need for Comprehensive Databases Effective multivariate analysis often requires extensive reference databases, which may not always be available for all food types or processing conditions (Esteki et al. 2019).

Influence of Processing Methods Food processing can alter the chemical or spectral profiles of products, potentially affecting the accuracy of multivariate models (Xia et al. 2019).

Variability in Raw Materials Natural variations in raw materials due to factors such as geographical origin or season can complicate the development of robust multivariate models (Cubero-Leon et al. 2014).

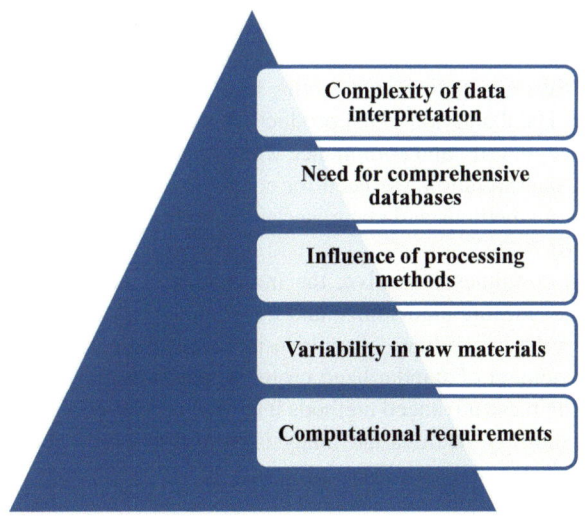

Fig. 3.3 Limitations of multivariate analysis in Halal food authentication

Table 3.2 Advances in multivariate analysis for Halal authentication

Advances in chemometrics	Description	References
Fusion of multiple analytical techniques	Combining data from different analytical methods (e.g., spectroscopy and chromatography) can provide more comprehensive authentication profiles	Borràs et al. (2015)
Machine learning and artificial intelligence	Advanced algorithms, such as artificial neural networks and support vector machines, are being applied to improve the accuracy and robustness of multivariate models	Bevilacqua et al. (2013)
Non-targeted metabolomics	Untargeted approaches that consider the entire metabolome of food products are emerging as powerful tools for detecting unknown adulterants	Rubert et al. (2016)
Portable multivariate systems	The development of field-deployable instruments capable of multivariate analysis can enhance on-site Halal authentication capabilities	Ellis et al. (2017)
Standardization of multivariate methods	Efforts to standardize multivariate analysis protocols across different laboratories and regions can improve consistency and reliability in Halal food authentication	Fathima et al. (2024)

Computational Requirements Some multivariate techniques require significant computational power, which may limit their applicability in routine or field testing scenarios (Cozzolino et al. 2011). To address these limitations and further improve multivariate analysis in Halal food authentication, several directions are being explored, as shown in Table 3.2.

Multivariate analysis has proven to be a powerful approach in the authentication of Halal food products, particularly for complex matrices such as meat, oils, fat, and fish. Its ability to consider multiple variables simultaneously provides a more comprehensive and robust authentication process compared to univariate methods. The success of multivariate analysis in detecting low levels of adulteration and differentiating between Halal and non-Halal products demonstrates its significant potential in ensuring food integrity and compliance with Halal standards. However, the complexity of data interpretation, the need for comprehensive databases, and the influence of processing methods and raw material variability present ongoing challenges Karahalil (2020).

As the field continues to evolve, the integration of advanced statistical techniques, machine learning algorithms, and novel analytical approaches promises to enhance further the capabilities of multivariate analysis in Halal food authentication. The development of standardized protocols and portable systems will be crucial in translating these advanced methods into practical, widely applicable tools for the Halal food industry. Multivariate analysis represents a critical component in the future of Halal food authentication.

3.2.2.1 Principle Component Analysis

Principal component analysis (PCA) is a popular unsupervised pattern recognition technique. PCA frequently serves as the initial phase in data analysis to find or validate patterns within observed data. It can reduce the dimensionality of complex data while maintaining the relationships between the data and enabling compression of the data set in order to visualize similarities and differences between chemically complex samples. PCA finds a set of orthogonal vectors that represents the data and then uses these vectors to model the data. By removing one or more of the least important vectors, it is possible to provide a much less dimensional representation of the data at the expense of only a small proportion of the information. By removing the first few components, the user can create a two- or three-dimensional representation of the data (Alahmadi et al. 2020; Hasan and Abdulazeez 2021; Taguchi 2024; Uddin et al. 2021; Bharadiya 2023; Ma et al. 2020).

PCA can categorize lard and other consumable fats and oils utilizing the plot of PC 1 and PC 2. Che Man et al. 2011 employed PCA to classify lard and 16 other edible fats and oils, using the absorbance of FTIR (spectra) at 16 distinct wavenumbers as variables. Approximately 90% of the variation is explained using four PCs (PC1, PC2, PC3, and PC4). PC 1 represented 44.1% of the variance, whilst PC 2 accounted for 30.2% of the variance. PCA effectively classified lard in pure ghee with FTIR spectral absorbances at optimal wavenumbers with 5% lard in pure ghee (Upadhaya et al. 2018). Likewise, bovine gelatin (Halal) and porcine gelatin (non-Halal) are categorized by PCA, utilizing characteristics derived from several analytical signals.

3.2.2.2 Cluster Analysis

Cluster analysis (CA), on the other hand, is an unsupervised pattern recognition technique that is used to classify objects in such a way that objects in the same cluster are more similar to each other than to objects in other clusters. CA results in the classification of a new object into one of the predefined clusters. Hierarchical cluster analysis (HCA) is a popular agglomerative hierarchical CA technique that is used to classify objects into a hierarchy of clusters, with a clear pictorial representation of the clusters produced at each step. If it is done that two clusters are neighboring to each other, they are combined into one bigger cluster (Zhang and Peng 2024; Naeem et al. 2023; Basar et al. 2020; Mehmood et al. 2022; Su et al. 2022; Ayaz et al. 2021; Elngar et al. 2021). Classification of lard, other animal fats and vegetable oils has been conducted using CA based on Mahalanobis distance, utilizing absorbance of FTIR spectra at 16 wavenumbers as variables (Che Man et al. 2011). It has also been shown that CA employing spectra at optimum specific frequency areas can be used to categorize fish, bovine, and porcine gelatins as well as their mixtures (Cebi et al. 2016).

3.2.3 Machine Learning and Artificial Intelligence

The application of machine learning (ML) and artificial intelligence (AI) in food authentication has revolutionized the field of Halal food verification. These advanced computational techniques offer powerful tools for analyzing complex data sets, recognizing patterns, and making predictions, which are particularly valuable in authenticating Halal meat, oils, fats, and fish products.

ML and AI techniques in Halal food authentication typically involve the development of predictive models based on large datasets generated from various analytical methods. These models can then be used to classify unknown samples or detect adulterants.

In meat authentication, ML and AI are often applied to spectroscopic or chromatographic data. Rady and Adedeji (2020) used Artificial Neural Networks (ANNs) to analyze hyperspectral imaging data for the detection of pork adulteration in minced beef. This method allows for rapid, non-destructive testing of meat products. Similarly, Kamruzzaman et al. (2013) employed Support Vector Machines (SVMs) to classify different meat species based on Near-Infrared (NIR) spectroscopy data. This approach demonstrates high accuracy in differentiating between Halal and non-Halal meat species.

For oils and fats, ML and AI techniques are frequently used to analyze complex chemical profiles. Jiménez-Carvelo et al. (2019) utilized Random Forests (RF) and ANNs to detect olive oil adulteration based on fatty acid profiles obtained through GC. This method can identify even subtle adulterations in vegetable oils. In another study, Georgouli et al. (2017) applied Deep Learning techniques to FTIR spectroscopy data for the detection of adulteration in extra virgin olive oil. The use of deep neural networks allowed for improved accuracy in identifying non-Halal additives.

Fish authentication often involves the analysis of genetic or protein data using ML and AI. Rather et al. (2024) reported the use of AI algorithms to analyze DNA metabarcoding data for the identification of fish species in complex food products. This approach allows for the simultaneous detection of multiple species, enhancing the ability to identify fish in processed foods. Politikos et al. (2021) employed Convolutional Neural Networks (CNNs) to analyze fish otolith images for species age identification. While primarily developed for ecological studies, this technique has potential applications in Halal fish authentication. ML and AI techniques have demonstrated remarkable success in Halal food authentication across various food categories. Despite its effectiveness, ML and AI approaches in Halal food authentication face several limitations, as shown in Fig. 3.4 and elaborated in the text.

Data Quality and Quantity ML and AI models require large, high-quality datasets for training, which may not always be available for all food types or processing conditions (Esteki et al. 2022).

Model Interpretability Some ML models, particularly deep learning networks, can be "black boxes," making it difficult to understand the basis of their predictions (Castelvecchi 2016).

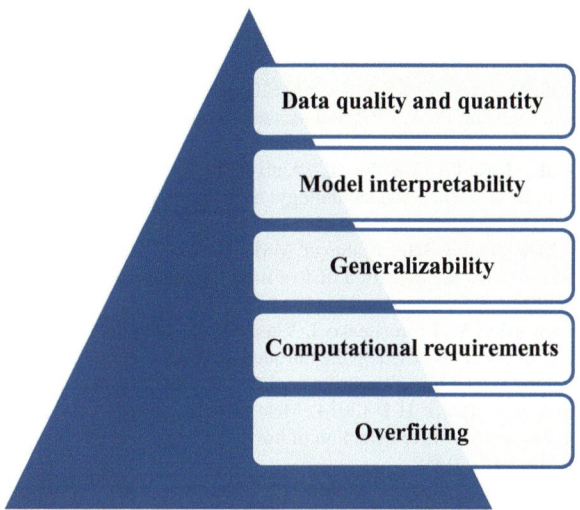

Fig. 3.4 Limitations of ML and AI analysis in Halal food authentication

Generalizability Models trained on specific datasets may not perform well when applied to samples from different geographical regions or processing conditions (Callao and Ruisánchez 2018).

Computational Requirements Advanced ML and AI techniques often require significant computational resources, which may limit their applicability in routine or field testing scenarios (Biancolillo and Marini 2018).

Overfitting There's a risk of models becoming too specialized to the training data, leading to poor performance on new, unseen samples (Gómez-Caravaca et al. 2016).

The success of ML and AI in detecting low levels of adulteration, differentiating between Halal and non-Halal products, and identifying species in complex mixtures demonstrates their significant potential in ensuring food integrity and compliance with Halal standards. However, challenges related to data quality and quantity, model interpretability, and generalizability present ongoing areas for improvement (Hassoun et al. 2020). As the field continues to evolve, the development of more sophisticated ML algorithms, the integration of AI with other analytical techniques, and the focus on explainable and generalizable models promise to further enhance the capabilities of these technologies in Halal food authentication. The advancement of edge computing and continuous learning models will be crucial in translating these advanced methods into practical, widely applicable tools for the Halal food industry (Hassoun et al. 2023).

References

Alahmadi A, Hussain M, Aboalsamh HA, Zuair M (2020) PCAPooL: unsupervised feature learning for face recognition using PCA, LBP, and pyramid pooling. Pattern Anal Applic 23:673–682. academia.edu

Aleixandre-Tudó JL, Castelló-Cogollos L, Aleixandre JL, Aleixandre-Benavent R (2022) Chemometrics in food science and technology: a bibliometric study. Chemom Intell Lab Syst 222:104514. sciencedirect.com

Ali ME, Razzak MA, Hamid SBA, Rahman MM, Al Amin M, Rashid NRA, Asing (2015) Multiplex PCR assay for the detection of five meat species forbidden in Islamic foods. Food Chem 177:214–224

Ayaz AM, Gorokhovatskyi V, Tvoroshenko I, Vlasenko N, Khalid MS (2021) The research of image classification methods based on the introducing cluster representation parameters for the structural description. Int J Eng Trends Technol 69(10):186–192. nure.ua

Bargen C, Brockmeyer J, Humpf H-U (2014) Meat authentication: a new HPLC–MS/MS based method for the fast and sensitive detection of horse and pork in highly processed food. J Agric Food Chem 62(39):9428–9435

Basar S, Ali M, Ochoa-Ruiz G, Zareei M, Waheed A, Adnan A (2020) Unsupervised color image segmentation: a case of RGB histogram based K-means clustering initialization. PLoS One 15(10):e0240015. plos.org

Bevilacqua M, Bucci R, Magrì AD, Magrì AL, Nescatelli R, Marini F (2013) Classification and class-modelling. In: Data handling in science and technology, vol 28. Elsevier, pp 171–233

Bharadiya JP (2023) A tutorial on principal component analysis for dimensionality reduction in machine learning. Int J Innov Sci Res Technol 8(5):2028–2032. researchgate.net

Bhat SA, Huang NF, Sofi IB, Sultan M (2022) Agriculture-food supply chain management based on blockchain and IoT: a narrative on enterprise blockchain interoperability. Agriculture 2022(12):40. https://doi.org/10.3390/agriculture12010040. mdpi.com

Biancolillo A, Marini F (2018) Chemometric methods for spectroscopy-based pharmaceutical analysis. Front Chem 6:576. https://doi.org/10.3389/fchem.2018.00576

Borràs E, Mestres M, Aceña L, Busto O, Ferré J, Boqué R, Calvo A (2015) Identification of olive oil sensory defects by multivariate analysis of mid infrared spectra. Food Chem 187:197–203

Callao MP, Ruisánchez I (2018) An overview of multivariate qualitative methods for food fraud detection. Food Control 86:283–293

Castelvecchi D (2016) Can we open the black box of AI? Nat News 538(7623):20

Cebi N, Durak MZ, Toker OS, Sagdic O, Arici M (2016) An evaluation of Fourier transforms infrared spectroscopy method for the classification and discrimination of bovine, porcine and fish gelatins. Food Chem 190:1109–1115

Chapman J, Truong VK, Elbourne A, Gangadoo S, Cheeseman S, Rajapaksha P et al (2020) Combining chemometrics and sensors: toward new applications in monitoring and environmental analysis. Chem Rev 120(13):6048–6069. [HTML]

Che Man YB, Rohman A, Mansor TST (2011) Differentiation of lard from other edible fats and oils by means of Fourier transform infrared spectroscopy and chemometrics. J Am Oil Chem Soc 88(2):187–192

Cozzolino D, Cynkar WU, Shah N, Smith P (2011) Multivariate data analysis applied to spectroscopy: potential application to juice and fruit quality. Food Res Int 44(7:1888–1896

Cubero-Leon E, Peñalver R, Maquet A (2014) Review on metabolomics for food authentication. Food Res Int 60:95–107

Ellis DI, Eccles R, Xu Y, Griffen J, Muhamadali H, Matousek PG, I. & Goodacre R. (2017) Through-container, extremely low concentration detection of multiple chemical markers of counterfeit alcohol using a handheld SORS device. Sci Rep 7:12082. https://doi.org/10.1038/s41598-017-12263-0

Elngar AA, Arafa M, Fathy A, Moustafa B, Mahmoud O, Shaban M, Fawzy N (2021) Image classification based on CNN: a survey. J Cybersecur Inf Manag 6(1):18–50. academia.edu

Esteki M, Regueiro J, Simal-Gandara J (2019) Tackling fraudsters with global strategies to expose fraud in the food chain. Compr Rev Food Sci Food Saf 18(2):425–440

Esteki M, Memarbashi N, Simal-Gandara J (2022) Classification and authentication of tea according to their geographical origin based on FT-IR fingerprinting using pattern recognition methods. J Food Compos Anal 106:104321

Fadzlillah NA, Rohman A, Ismail A, Mustafa S, Khatib A (2013) Application of FTIR-ATR spectroscopy coupled with multivariate analysis for rapid estimation of butter adulteration. J Oleo Sci 62(8):555–562

Fathima AM, Rahmawati L, Windarsih A, Suratno S (2024) Advanced Halal authentication methods and technology for addressing non-compliance concerns in Halal meat and meat products supply chain: a review. Food Sci Anim Resour 44(6):1195–1212

Ferone M, Gowen A, Fanning S, Scannell AG (2020) Microbial detection and identification methods: bench top assays to omics approaches. Compr Rev Food Sci Food Saf 19(6):3106–3129. researchgate.net

Georgiou CA, Danezis GP (2015) Elemental and isotopic mass spectrometry. In: Pico Y (ed) Comprehensive analytical chemistry, vol 68., Chapter 3. Amsterdam, Elsevier, pp 131–243

Georgouli K, Del Rincon JM, Koidis A (2017) Continuous statistical modelling for rapid detection of adulteration of extra virgin olive oil using mid infrared and Raman spectroscopic data. Food Chem 217:735–742

Gómez-Caravaca AM, Maggio RM, Cerretani L (2016) Chemometric applications to assess quality and critical parameters of virgin and extra-virgin olive oil: a review. Anal Chim Acta 913:1–21. https://doi.org/10.1016/j.aca.2016.01.025

Grundy HH, Reece P, Buckley M, Solazzo CM, Dowle AA, Ashford D et al (2016) A mass spectrometry method for the determination of the species of origin of gelatine in foods and pharmaceutical products. Food Chem 190:276–284

Guntarti A, Martono S, Yuswanto A, Rohman A (2017) Analysis of beef meatball adulteration with wild boar meat using real-time polymerase chain reaction. Int Food Res J 24(6):2451–2455

Hafis Yuswan M, Shirwan Abdullah Sani M, Noorzianna Abdul Manaf Y, Nasir Mohd Desa M (2020) Basic requirements of laboratory operation for Halal analysis. KnE Soc Sci 4(9):55–65

Hasan BMS, Abdulazeez AM (2021) A review of principal component analysis algorithm for dimensionality reduction. J Soft Comput Data Min 2(1):20–30. uthm.edu.my

Hassoun A, Måge I, Schmidt WF, Temiz HT, Li L, Kim HY et al (2020) Fraud in animal origin food products: advances in emerging spectroscopic detection methods over the past five years. Foods 9(8):1069

Hassoun A, Aït-Kaddour A, Abu-Mahfouz AM, Rathod NB, Bader F, Barba FJ et al (2023) The fourth industrial revolution in the food industry—part I: industry 4.0 technologies. Crit Rev Food Sci Nutr 63(23):6547–6563. unit.no

Heidari M, Talebpour Z, Abdollahpour Z, Adib N, Ghanavi Z, Aboul-Enein HY (2020) Discrimination between vegetable oil and animal fat by a metabolomics approach using gas chromatography-mass spectrometry combined with chemometrics. J Food Sci Technol 57(9):3415–3425. https://doi.org/10.1007/s13197-020-04375-9

Jiménez-Carvelo AM, González-Casado A, Bagur-González MG, Cuadros-Rodríguez L (2019) Alternative data mining/machine learning methods for the analytical evaluation of food quality and authenticity–a review. Food Res Int 122:25–39

Kamruzzaman M, Sun DW, ElMasry G, Allen P (2013) Fast detection and visualization of minced lamb meat adulteration using NIR hyperspectral imaging and multivariate image analysis. Talanta 103:130–136

Karahalil E (2020) Principles of Halal-compliant fermentations: microbial alternatives for the Halal food industry. Trends Food Sci Technol 98:1–9

Kharbach M, Alaoui Mansouri M, Taabouz M, Yu H (2023) Current application of advancing spectroscopy techniques in food analysis: data handling with chemometric approaches. Foods 12:2753. mdpi.com

Liu GL, Kazarian SG (2022) Recent advances and applications to cultural heritage using ATR-FTIR spectroscopy and ATR-FTIR spectroscopic imaging. Analyst 147:1777–1797. rsc.org

Liu Y, Pu H, Sun DW (2021) Efficient extraction of deep image features using convolutional neural network (CNN) for applications in detecting and analysing complex food matrices. Trends Food Sci Technol 113:193–204. [HTML]

Liu C, Zuo Z, Xu F, Wang Y (2023) Authentication of herbal medicines based on modern analytical technology combined with chemometrics approach: a review. Crit Rev Anal Chem 53(7):1393–1418. [HTML]

Ma Z, Liu Z, Zhao Y, Zhang L, Liu D, Ren T et al (2020) An unsupervised crop classification method based on principal components isometric binning. ISPRS Int J Geo-Inf 9(11):648. mdpi.com

Marikkar J, Ghazali H, Man YC, Peiris T, Lai O (2005) Distinguishing lard from other animal fats in admixtures of some vegetable oils using liquid chromatographic data coupled with multivariate data analysis. Food Chem 91(1):5–14

Marikkar JMN, Mirghani MES, Jaswir I (2016) Application of chromatographic and infra-red spectroscopic techniques for detection of adulteration in food lipids: a review. J Food Chem Nanotechnol 2(1):32–41

Martinez I, Jakobsen Friis T (2004) Application of proteome analysis to seafood authentication. Proteomics 4(2):347–354

Mehmood M, Shahzad A, Zafar B, Shabbir A, Ali N (2022) Remote sensing image classification: a comprehensive review and applications. Math Probl Eng 2022(1):5880959. wiley.com

Mottola A, Piredda R, Catanese G, Giorelli F, Cagnazzo G, Ciccarese G, Dambrosio A, Quaglia NC, Di Pinto A (2022) DNA metabarcoding for identification of species used in fish burgers. Ital J Food Saf 11(3):10412. https://doi.org/10.4081/ijfs.2022.10412

Naeem S, Ali A, Anam S, Ahmed MM (2023) An unsupervised machine learning algorithms: comprehensive review. Int J Comput Digit Syst 13(1):911–921. https://doi.org/10.12785/ijcds/130172. researchgate.net

Otto M (2023) Chemometrics: statistics and computer application in analytical chemistry, 4th edn ISBN: 978-3-527-84379-4. psu.edu

Politikos DV, Petasis G, Chatzispyrou A, Mytilineou C, Anastasopoulou A (2021) Automating fish age estimation combining otolith images and deep learning: the role of multitask learning. Fish Res 242:106033

Rady A, Adedeji AA (2020) Application of hyperspectral imaging and machine learning methods to detect and quantify adulterants in minced meats. Food Anal Methods 13:970–981. https://doi.org/10.1007/s12161-020-01719-1

Rather MA, Ahmad I, Shah A, Hajam YA, Amin A, Khursheed S, Ahmad I, Rasool S (2024) Exploring opportunities of artificial intelligence in aquaculture to meet increasing food demand. Food Chem X 22(2024):101309

Rizou M, Galanakis IM, Aldawoud TM, Galanakis CM (2020) Safety of foods, food supply chain and environment within the COVID-19 pandemic. Trends Food Sci Technol 102:293–299. nih.gov

Rohman A, Che Man YB (2011) The use of Fourier transform mid infrared (FT-MIR) spectroscopy for detection and quantification of adulteration in virgin coconut oil. Food Chem 129(2):583–588

Rohman A, Erwanto Y, Man YBC (2011) Analysis of pork adulteration in beef meatball using Fourier transform infrared (FTIR) spectroscopy. Meat Sci 88(1):91–95

Rohman A, Ghazali MAIB, Windarsih A, Riyanto S, Yusof FM, Mustafa S (2020) Comprehensive review on application of FTIR spectroscopy coupled with chemometrics for authentication analysis of fats and oils in the food products. Molecules 25(22):5485

Rohman A, Orbayinah S, Hermawan A, Sudjadi S, Windarsih A, Handayani S (2022) The development of real-time polymerase chain reaction for identification of beef meatball. Appl Food Res 2(2):100148

Rubert J et al (2016) Anal Bioanal Chem 408(14):3657–3667

Scarano D, Rao R (2014) DNA markers for food products authentication. Diversity 6(3):579–596

Shehata HR, Naaum AM, Garduño RA, Hanner R (2018) DNA barcoding as a regulatory tool for seafood authentication in Canada. Food Control 92:147–153

Shen Y, Xu L, Li Y (2021) Biosensors for rapid detection of salmonella in food: a review. Compr Rev Food Sci Food Saf 20(1):149–197. [HTML]

Su Y, Gao L, Jiang M, Plaza A, Sun X, Zhang B (2022) NSCKL: normalized spectral clustering with kernel-based learning for semisupervised hyperspectral image classification. IEEE Trans Cybern 53(10):6649–6662. [HTML]

Swarbrick B, Westad F (2016) An overview of chemometrics for the engineering and measurement sciences. In: Handbook of measurement in science and engineering. Wiley, Hoboken, p 2309

Taguchi Y (2024) Unsupervised feature extraction applied to bioinformatics: a PCA based and TD based approach, 2nd edn. Springer, Cham. https://doi.org/10.1007/978-3-031-60982-4. researchgate.net

Todorov H, Searle-White E, Gerber S (2020) Applying univariate vs. multivariate statistics to investigate therapeutic efficacy in (pre)clinical trials: a Monte Carlo simulation study on the example of a controlled preclinical neurotrauma trial. PLoS One 15(3):e0230798

Tortorella S, Cinti S (2021) How can chemometrics support the development of point of need devices? Anal Chem 93:2713–2722. acs.org

Uddin MP, Mamun MA, Afjal MI, Hossain MA (2021) Information-theoretic feature selection with segmentation-based folded principal component analysis (PCA) for hyperspectral image classification. Int J Remote Sens 42(1):286–321. [HTML]

Upadhaya N, Jaiswal P, Jha SN (2018) Application of attenuated total reflectance Fourier transform infrared spectroscopy (ATR-FTIR) in MIR range coupled with chemometrics for detection of pic body fat in pure ghee (heat clarified milk fat). J Mol Struct 1153:275–281

Wang HP, Chen P, Dai JW, Liu D, Li JY, Xu YP, Chu XL (2022) Recent advances of chemometric calibration methods in modern spectroscopy: algorithms, strategy, and related issues. TrAC Trends Anal Chem 153:116648. [HTML]

Xia J, Zhang J, Zhao Y, Huang Y, Xiong Y, Min S (2019) Fourier transform infrared spectroscopy and chemometrics for the discrimination of paper relic types. Spectrochim Acta A Mol Biomol Spectrosc 219:8–14

Zhang H, Peng Y (2024) Image clustering: an unsupervised approach to categorize visual data in social science research. Sociol Methods Res 53(3):1534–1587. [HTML]

Zhang X, Yang J (2024) Advanced chemometrics toward robust spectral analysis for fruit quality evaluation. Trends Food Sci Technol 150:104612. [HTML]

Zhang Y, Deng L, Zhu H, Wang W, Ren Z, Zhou Q et al (2023) Deep learning in food category recognition. Inf Fusion 98:101859. sciencedirect.com

Chapter 4
Halal Food Authentication Using FTIR Spectroscopy

4.1 Introduction

Food authentication has emerged as a pressing concern, sparking significant attention from consumers, regulatory authorities, and the food industry. The Food and Drugs Authority (FDA), Food and Agriculture Organization (FAO), Codex Alimentarius (CODEX), and World Health Organization (WHO) have all intervened to ensure food safety, recognizing the critical role that food authentication plays in protecting public health. Food authentication is intricately linked to food security, which the FAO defines as "a state in which everyone, everywhere, has physical and financial access to enough wholesome food that satisfies their dietary requirements and food choices for a healthy and active lifestyle" (FAO 2014). Food authenticity is a complex issue influenced by factors such as legal compliance, religious needs like Halal and Kosher standards, economic aspects, quality control, and safety. The replacement of high-cost ingredients with cheaper substitutes is a common practice that affects food authenticity. For example, pig-derived products are often more affordable than Halal-certified alternatives, making it vital to ensure the Halal status of ingredients (Popping et al. 2022). This is especially important for Muslim consumers who rely on Halal-certified products to meet their dietary guidelines. To tackle these challenges, reliable techniques and effective detection methods are needed for Halal verification. Various approaches are employed to detect and quantify adulterants in food systems, as shown in Table 4.1.

Table 4.1 Approaches used to detect and quantify adulterants in food authentication

Approach	Brief description of approaches
First approach	Determines the ratios of specific chemical constituents, assuming these ratios remain constant in pure food. Any deviation indicates adulteration or anomalies in the chemical composition
Second approach	Identifies specific markers, such as chemical constituents or morphological components that signal the presence of adulterants
Third approach	Uses analytical methods based on the physicochemical properties of the specific food components

4.2 Importance of Halal Certification

Halal certification plays a vital role in the global food and non-food industries, driven by the increasing demand for Halal products from both Muslim and non-Muslim consumers. Halal certification is a procedure that ensures a product or service adheres to the principles and guidelines of Islamic law (Anggarkasih and Resma 2022). The certification process generally includes an assessment of the production methods, ingredients, and facilities to confirm compliance with Halal requirements. The scope of Halal certification encompasses not only food products but also non-food items, services, and even tourism destinations (Abdul Latif et al. 2014). Halal certification can be understood from several perspectives, as shown in Fig. 4.1.

4.3 Foods That Need Halal Authentication

With the widespread availability of Halal foods (Fig. 4.2), both Muslims and non-Muslims are increasingly concerned about the authenticity of Halal food items (Kurniadi and Frediansyah 2017). All forms of food [meat and poultry, processed meat products, processed seafood, dairy products, edible oils, bakery products, confectionary (gelatin-based products), beverages, and nutritional products (pharmaceuticals)], especially meat items, must have Halal certification. This can be attributed to the fact that meat products are essential components of human diets. Consumption of food and pharmaceuticals is vital to human survival (Ng et al. 2022). These items must satisfy consumer demand in addition to biochemistry, globalization, and industrialization technology advancement. Innovative technology has significantly enhanced the domain of food (science and technology). However, certain producers benefit from this development by including non-Halal ingredients, thus engaging in fraudulent practices against their target consumers. Some food products are misrepresented regarding the origin of their ingredients (Montowska and Pospiech 2010), and some packaged food is not Halal certified.

4.3 Foods That Need Halal Authentication

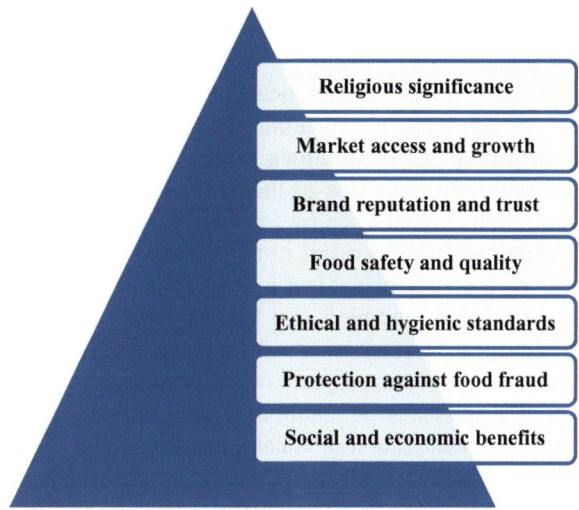

Fig. 4.1 Common points in Halal certification

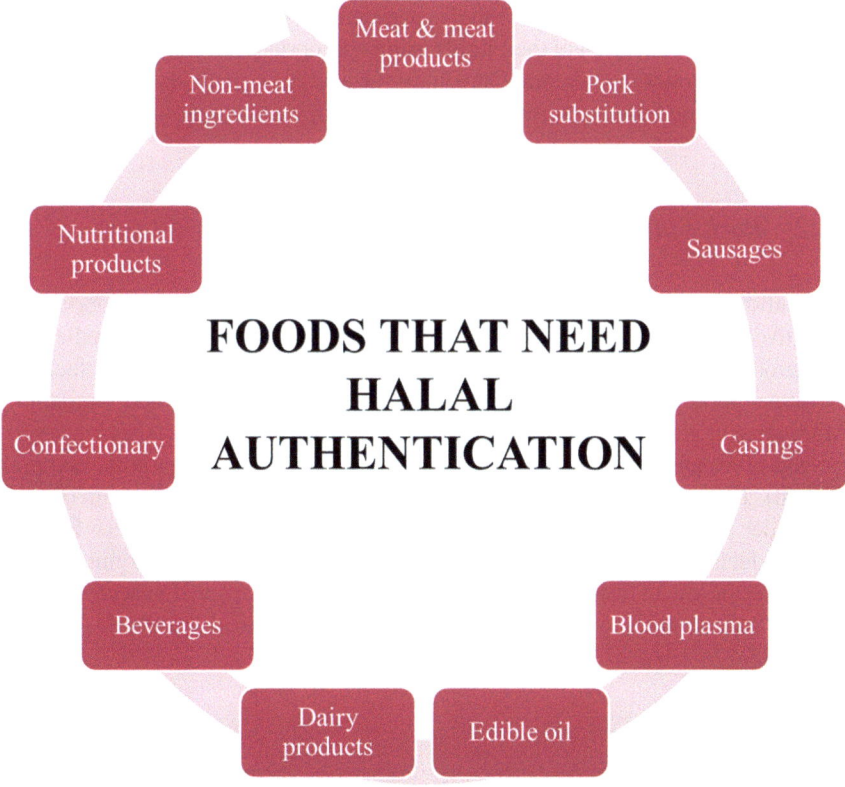

Fig. 4.2 Halal compliance in commonly consumed foods

4.3.1 Meat and Meat Products

Increasing demand and expenses render meat and meat-related items more susceptible to unauthorized adulteration, substitution, and deceptive marketing practices. For consumers with religious prohibitions against pork eating, counterfeit meat products pose a significant issue due to their affordability (Hassan et al. 2021). Muslim customers are frustrated when beef meat is mixed with or replaced by pork meat; this issue is particularly prevalent in non-Muslim nations, complicating the acquisition of Halal-certified meals. Meat is a significant issue in relation to Halal food; the meat from animals and birds that Muslims are authorized to consume is referred to as Halal meat in accordance with Islamic standards (Candoğan et al. 2021). Diverse approaches are employed to distinguish pork meat from other meat products. The price disparity is an additional factor prompting manufacturers to engage in the unlawful adulteration of Halal meat (beef) with non-Halal meat (rat meat). The industrialization of food manufacturing unveils Muslims with exposure to non-Halal ingredients, such as blood plasma, gelatin, and transglutaminase, found in meatballs, frankfurters, and surimi products (Rohman 2018). The key challenge in the meat sector is verifying food authenticity, as Halal meat products are frequently tainted with haram substances (Ahmad and Ayub 2022). Meat-based products that do not comply with Halal criteria are subjected to instrumental analysis following the extraction of their lipids using appropriate solvents and techniques. Figure 4.3 illustrates the schematic layout of the analytical methods for the analysis of meat-based food products using FTIR spectroscopy and chemometrics.

4.3.2 Pork Substitution

Pork substitution is a fraudulent practice in the food industry where pork or its derivatives are used as substitutes for more expensive or authentic ingredients in food products. This issue arises because, in many countries, food manufacturers substitute pork products in food items due to their low cost and availability (Murugaiah et al. 2009). These pork derivatives include pork tissues such as collagen and offal, mechanically recovered meats (MRM) derived from pork, and pork fat (lard). The use of such ingredients in food products renders them haram (forbidden) under Islamic law. Collagen and offal are often used as inexpensive fillers in meat products to reduce production costs while maintaining profit margins. If these ingredients originate from pigs, the resulting products are deemed unacceptable for Muslim consumption, violating both religious and ethical standards (Saputra et al. 2018).

4.3 Foods That Need Halal Authentication

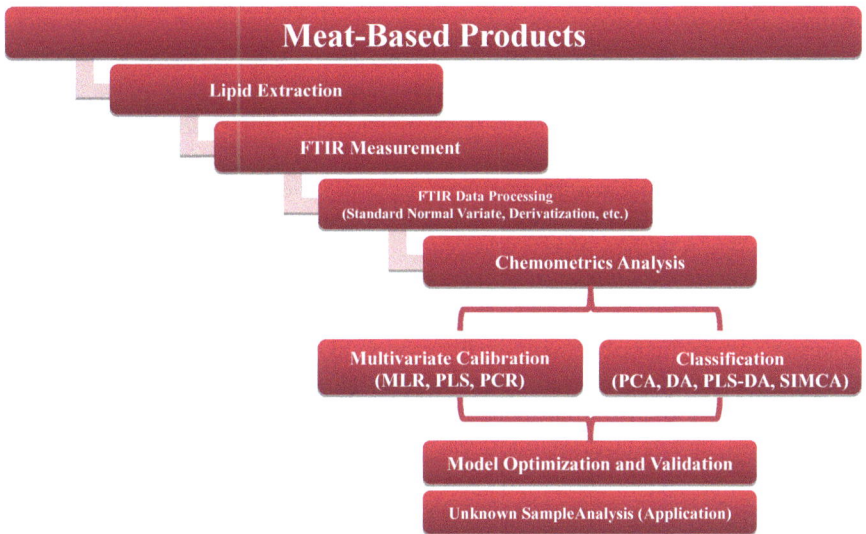

Fig. 4.3 Analytical procedure for analysis of meat-based food products using FTIR spectroscopic method

Pork fat, or lard, is another commonly used substitute in meat products. It is often used fraudulently to replace more expensive animal fats, such as those derived from cows or sheep. This substitution not only compromises the Halal integrity of the product but also undermines consumer trust, particularly in regions where Halal certification is expected (Guntarti et al. 2020). MRM presents another significant challenge in ensuring Halal compliance. MRM is a paste-like meat product obtained by mechanically separating residual meat from bones under high pressure. While MRM offers the food industry a cost-effective way to utilize otherwise wasted meat, its use of pork carcasses for production makes the resulting products haram (Surowiec et al. 2011). Commonly incorporated into comminuted meat-based products such as sausages, meat pies, and burgers, MRM is widely used due to its high calcium and iron content and low collagen levels. However, when derived from pork carcasses, these products are forbidden for Muslim consumption (Syofyan et al. 2025). The fraudulent use of pork-derived MRM and other substitutes in Halal-labeled products represents a serious violation of religious dietary laws. This practice not only poses ethical concerns but also highlights the economic motivations driving food fraud. Regulatory bodies and food manufacturers must implement rigorous controls and transparent processes to prevent such fraudulent practices and safeguard consumer trust (Van Mierlo et al. 2022).

4.3.3 Sausages

Sausage is a meat product prepared by filling ground meat, typically blended with spices, salt, and herbs, into a casing made either from natural intestine or synthetic materials. Sausage is a widely loved and esteemed meat product globally. Sausages can be made from chicken, mutton, beef, or pork. The ancient Chinese produced sausages from pork, leading to the myth that Muslims abstain from consuming sausages (Flores and Piornos 2021). Nevertheless, sausages (beef) are quite common in Muslim countries. Soujouk is a specific name for a variety of products that have been made for a long time and are popular in Turkiye and other Middle Eastern countries. Another pure beef sausage that is well-liked in many Muslim countries is "Merguez", which is produced from beef wrapped in sheep casing. In contrast to pork sausages, which are considered haram, sausages made from chicken, mutton, and beef that are filled with cellulose casings or animal (goat cattle and sheep) casings sourced from Halal-slaughtered animals are considered Halal (Ramos-Moreno et al. 2021).

4.3.4 Casings

Casings are mostly utilized to ascertain dimensions and provide appearance to meat products, particularly sausages. They serve as processing molds, principal packaging and shipping, and marketing units during display and are derived from cellulose and collagen (Ledesma et al. 2016). There are four specific types of casing commonly utilized, as shown in Fig. 4.4. Cellulose casings are deemed Halal as they are derived from plant materials. Animal casings, on the other hand, are derived from the intestines of various animals. The intestines can be obtained from goats and sheep, as well as pigs, which are good sources for the intestine (Henchion et al. 2017). Casings derived from sheep or goats are considered Halal, whereas casing from pigs is considered haram and consequently prohibited for Muslim consumption. Sheep and goat casings are considered Halal only if the animals are slaughtered in accordance with Halal practices. Similarly, collagen casings (finely ground animal skin) are also edible if obtained from Halal slaughtered animals (Rivas et al. 2018).

4.3.5 Blood Plasma

Blood plasma is frequently encompassed in meat products because of its gelling and emulsifying characteristics. Porcine blood and its byproducts, including plasma and red blood cells, are being used by the food industry as food ingredients (Herrero

4.3 Foods That Need Halal Authentication

Fig. 4.4 Specific types of casings used in industrial product

et al. 2009). These products are typically marketed as spray-dried powders due to their high biological value and excellent functional characteristics. The utilization of blood plasma, regardless of its origin, is deemed haram and hence forbidden for Muslim consumption. It is, therefore, unacceptable for Muslim consumers to consume any product that contains blood (Wang et al. 2022a, b).

4.3.6 Non-meat Ingredients

Various organic and synthetic compounds are often added to meat products as colorants, flavor enhancers, preservatives, aromas, thickeners, binders, or stabilizers. It is crucial to ensure that prohibited substances are not used in Halal meat products (Owusu-Ansah et al. 2022). Some of the most common non-Halal ingredients include gelatine, which is classified as a food ingredient under the EEC's Codex Alimentarius and is typically derived from animals unless specifically labeled as "Halal gelatine". Other non-Halal additives include glycerine and lecithin derived from animal fats, as well as alcohol (El Sheikha et al. 2017). Additionally, certain pork-derived ingredients, such as lard, mono and diglycerides, sodium stearoyl lactylate, and polysorbate 60 or 80, are commonly used in food production. Enzymes from non-Halal animals, grain or plant-based ingredients with pork-derived carriers such as β-carotene (pig Gelatin) and BHA/BHT with a pig-based carrier are also considered non-compliant with Halal standards. Other examples include blood plasma enzymes, bacon, or natural bacon flavor (Eckl et al. 2021). Some ingredients are classified as doubtful, including yeast extract from brewer's yeast and cochineal/

carmine color, which should also be avoided. To eliminate any uncertainty regarding the Halal status of ingredients, meat processors are encouraged to request Halal certification from their suppliers for all food additives and processing aids (Nunes et al. 2016).

4.3.7 Beverages or Drinks

Alcohol, specifically ethyl alcohol, is commonly used in food production or generated during processing. Given its widespread applications in the pharmaceutical, food, and cosmetic industries, the Halal status of alcohol warrants attention. The presence of ethanol in food products is a contentious issue. Islamic jurists consider ethanol to be haram. However, some beverages contain trace amounts of alcohol, necessitating accurate identification (Wang et al. 2017; Karim and Muhamad 2018).

4.4 Perception of Food Fraud and Adulteration

Food adulteration is the intentional manipulation of food quality by incorporating additional components to modify various attributes of food products, including but not limited to appearance, taste, color, volume, or weight. Food fraud is adulterating food for economic reasons, commonly observed in most circumstances (Kendall et al. 2019). This morally objectionable behavior entails a possible hazard to human well-being, either through a brief exposure that may lead to immediate toxicity or through prolonged interaction that may result in enduring impacts that are challenging to evaluate (Kowalska 2018). Adulteration and fraud of foods encompass three primary categories of fraudulent economic practices:

1. Addition of Cheaper Substitutes or Non-Nutritive Components.
2. Substitution with Inferior Quality Products.
3. Misrepresentation of Food Origin or Production Method.

The issue of food security has become increasingly prominent in the twenty-first century, underscoring the necessity for effective international systems to prevent and mitigate food fraud within global food supply chains. Successful management of food fraud necessitates implementing intelligent systems that primarily rely on preventive measures, proactive inspection, and techniques for mitigation. The recognition of state-of-the-art analytical competence is crucial in verifying and validating preventative, proactive, transparent, and systematic food quality and safety programs (Djekic and Smigic 2023; Soon-Sinclair et al. 2024).

4.5 Challenges in Halal Food Authentication

Food authentication is a complex process that faces several challenges, as shown in Fig. 4.5, and each is elaborated on in the text one by one (Ng et al. 2022; Danezis et al. 2016).

Food adulteration: Food adulteration refers to the deliberate incorporation of a substance into a food product to enhance its market value or profitability. Food adulteration can involve the addition of a cheaper ingredient to a food product, such as the addition of lard in palm oil or melamine to milk powder.

Mislabeling: Mislabeling refers to the fraudulent labeling of a food product, exemplified by the classification of a food product as "organic" when it doesn't meet that criterion.

Counterfeiting: Counterfeiting refers to the manufacture and distribution of fake or imitation food items, including the development and sale of counterfeit olive oil or honey.

Complexity of food supply chains: Food supply chains are complex and encompass multiple stakeholders, including growers, processors, producers, suppliers, and retailers. This complexity can hinder the tracing of the origin and composition of foodstuffs.

Limited resources: Food authentication necessitates substantial resources, encompassing equipment, personnel, and financial support. Limited resources can hinder the efficacy of food verification efforts.

Rapidly changing food technology: Food technology is evolving swiftly, with the continuous development of novel ingredients and processing techniques. This

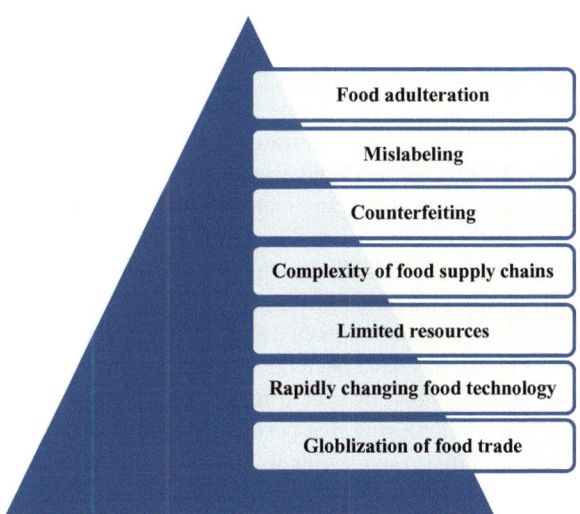

Fig. 4.5 Common challenges in food authentication

may hinder the ability to stay abreast of recent advancements in food authentication.

Globalization of food trade: The globalization of food trade has led to an increase in the importation and exportation of food products. This can make it difficult to track the origin and composition of food products.

4.6 Spectroscopic Techniques for Food Authentication

Various analytical methods are designed to exploit differences in the composition and nature of minor and major components present or absent in food. These methods often depend on physical-chemical constants or chemical and biological measurements. Ideally, these methods are effective for food authentication; however, they necessitate high technical expertise and can be costly, along with the need for reagents and waste management budgets (Biancolillo et al. 2020). Owing to these and additional challenges, including their cumbersome nature, prolonged analytical duration, and intricate instrument design, several researchers have concentrated on expeditious noninvasive techniques such as molecular spectroscopy, for instance, ultraviolet-visible spectroscopy (UV/Vis), infrared spectroscopy (IR), Raman spectroscopy (RS), nuclear magnetic resonance (NMR) spectroscopy, and hyperspectral imaging (HSI). Generally, these techniques have been recognized for use in both small-scale and industrial contexts across numerous fields. Widely used spectroscopic methods for food evaluation encompass IR, Raman, and NMR spectroscopy, each contributing to assessing food authenticity, safety, and quality (Karabagias 2020; Haider et al. 2024; Rohman and Windarsih 2020).

4.6.1 Theoretical Background of Infrared Spectroscopy

When it comes to analytical methods, infrared (IR) spectroscopy is among the most crucial tools at the disposal of modern scientists. The simple fact that IR spectroscopy is used to investigate virtually any material in almost every state is one of the most significant advantages of this technique. A reasonable selection of sample techniques allows for the examination of a wide variety of substances, including liquids, solutions, pastes, powders, films, fibers, gases, and surfaces. A range of new sensitive procedures have been established as a result of the improvement in instrumentation. These approaches have been developed to investigate previously intractable samples. Since the 1940s, IR spectrometers have been available for purchase on the commercial market. During that period, the instruments relied on prisms to perform the function of dispersive elements. However, by the middle of the 1950s, diffraction gratings had been incorporated into the technology of dispersive machines (Barbara Stuart 2004; Beć and Huck 2019). The pivotal progress in IR spectroscopy has been instigated by the creation of Fourier-transform spectrometers, significantly propelling the field forward. These spectrometers leverage Fourier transformation, a well-established mathematical method implemented by the instrument utilizing an interferometer. Thanks to the evolution of Fourier-transform infrared (FTIR) spectroscopy, there has been an enhancement in the quality of IR spectra, coupled with a notable reduction in the time required for data collection. Furthermore, ongoing advancements in computer technology have contributed to additional significant progress in IR. This technique operates on the principle of capturing the vibrations of a molecule's atoms. The procedure typically involves transmitting IR through a sample and quantifying the portion of incident radiation absorbed at a specific energy level to generate an IR. The frequency of vibration within a sample molecule corresponds to the energy level at which any peak emerges in an absorption spectrum (Barbara Stuart 2007; Dutta 2017).

4.6.2 Electromagnetic Radiation

In the electromagnetic spectrum, the visible portion is defined as the portion of the spectrum that contains radiation that the human eye can see. Radio waves, microwaves, infrared, ultraviolet, X-ray, and gamma-ray are the many types of radiation that are detected by other detection systems (Zwinkels 2015). These methods disclose radiation that extends beyond the visible parts of the spectrum. Table 4.2 depicts the regions and the processes of radiation contact with matter. Either classical or quantum theories can be used to explain the electromagnetic spectrum and the wide range of interactions that occur between the many types of radiation and the various types of matter.

4.6.3 Infrared Absorption

In order for a molecule to exhibit IR absorptions, the molecule must possess a particular characteristic; specifically, the molecule's electric dipole moment must change while it is vibrating. Within the realm of IR spectroscopy, this is the selection rule. A heteronuclear diatomic molecule, as shown in Fig. 4.6, is one type of

Table 4.2 Electromagnetic spectrum from gamma rays to radio waves

Type of radiation	Gamma rays	X-rays	Ultraviolet 10–400/380	Visible 380–700	Infrared				Radio waves
					Near	Mid	Far	Microwave	
Wavelength range	$<10^{-12}$ m–1 nm		1 nm–750 nm		750 nm–1000 µm			1 mm–25 µm	>1 mm
Frequency (Hz)	$10^{24}-10^{17}$		$10^{17}-10^{14}$		$10^{14}-10^{13}$			$10^{13}-10^{11}$	$<10^{11}$

Fig. 4.6 Change in the dipole moment of a heteronuclear diatomic molecule. (Barbara Stuart 2004)

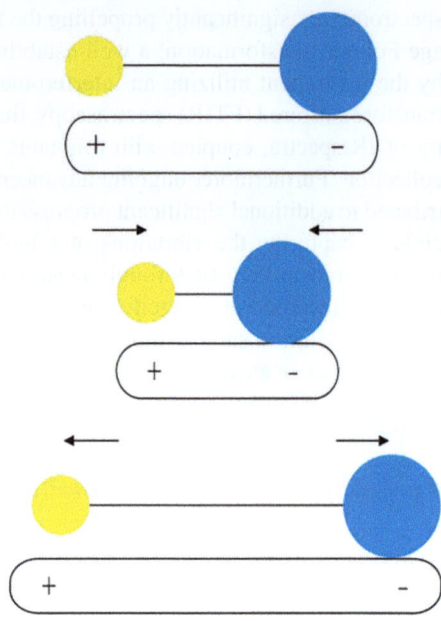

"IR-active" molecule. The molecule's dipole moment fluctuates as the bond stretches and compresses. In contrast, a homonuclear diatomic molecule is an example of an "IR-inactive" molecule since its dipole moment is always zero, regardless of the length of the bond. It's crucial to comprehend group theory and molecular symmetry when first assigning IR bands. Beyond the limits of this book, a full explanation of this theory is not possible. However, other texts dedicate a significant amount of space to discussing symmetry and group theory (Dutta 2017).

4.6.4 Normal Modes of Vibration

By utilizing the variations in molecular dipoles associated with vibrations and rotations, one may comprehend how IR light interacts with matter. One can start by envisioning a molecule as a collection of masses connected by bonds that display spring-like properties. An example that is considered to be elementary would be diatomic molecules. These molecules, in addition to having two degrees of rotational freedom, also have three degrees of translational freedom (Pavia et al. 2008). In addition, the atoms that make up the molecules are able to move in relation to one another. This means that the bond lengths can change, or an individual atom can move away from its present plane. The movements of stretching and bending are collectively referred to as vibrations, and this is a description of those movements taken together. When it comes to a diatomic molecule, only one vibration can occur, and it corresponds to the stretching and compression of the bond (Mikhlin and

4.6 Spectroscopic Techniques for Food Authentication

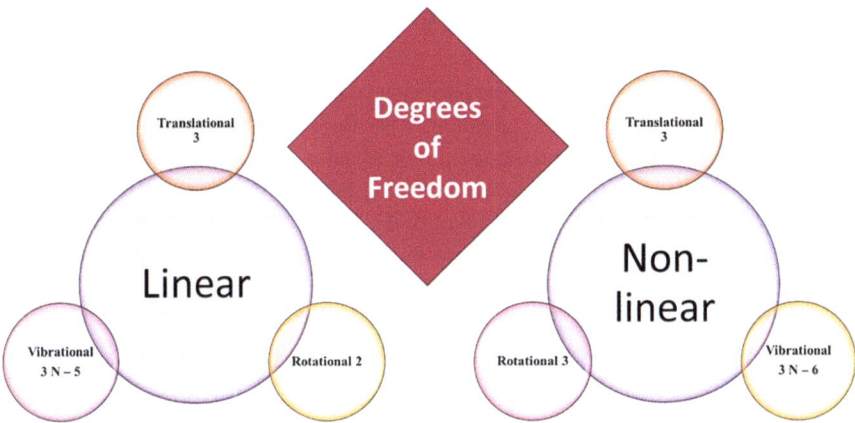

Fig. 4.7 Degrees of freedom for polyatomic molecules

Avramov 2015). This represents one degree of freedom with respect to vibrations. If a polyatomic molecule has (N) atoms, then it has 3 N degrees of freedom. Taking into consideration the case of molecules that include three atoms, it is possible to differentiate between two types of triatomic molecules, namely, linear and non-linear polyatomic molecules. Figure 4.7 shows the summarized degrees of freedom of polyatomic molecules.

4.6.5 Sources and Detectors

FTIR spectrometers typically utilize a Nernst or Globar source when operating in the mid-IR region. In cases where the examination extends to the far-IR region, a high-pressure mercury lamp becomes the preferred light source. For the near-IR region, a tungsten-halogen lamp is commonly chosen. Additionally, two frequently employed detectors are utilized for the mid-IR region. The standard detector widely used for regular applications consists of a pyroelectric device with deuterium triglycine sulfate (DTGS) enclosed in a temperature-resistant alkali halide window. Mercury cadmium telluride (MCT) can be employed when heightened sensitivity is required, although it necessitates cooling to liquid nitrogen temperatures (Barbara Stuart 2004). In the far-IR spectrum, germanium or indium–antimony detectors are utilized, functioning at liquid helium temperatures. In the near-IR range, the typically employed detectors are lead sulfide photoconductors (Chu et al. 2024).

4.6.6 Advantages of Fourier-Transformation

When compared to previous dispersive instruments, FTIR devices have several major advantages. Two of these advantages are the Fellgett advantage, also known as the multiplex advantage, and the Jacquinot advantage, also known as the throughput advantage. Because of an increase in the signal-to-noise ratio (SNR) per unit of time, which is proportional to the square root of the number of resolution elements that are being monitored, the Fellgett advantage is beneficial. This is a consequence of the enormous number of resolution elements that are being monitored simultaneously. The complete source output can be continuously passed through the sample because FTIR spectrometry does not require the use of a slit or any other restricting device. This allows for the spectrometry to be performed without interruption. This leads to a significant increase in the amount of energy that is absorbed by the detector, which ultimately results in increased signal strength and improved SNR. This is referred to as Jacquinot's Advantage. The speed advantage that FTIR spectrometry offers is yet another one of its strengths. Spectra can be obtained in a millisecond due to the mirror's capacity to move short distances relatively quickly. This ability, in conjunction with the gains in SNR that are a result of the Fellgett and Jacquinot advantages, makes it possible to collect spectra. One of the factors that determines the precision of the position of an IR band in interferometry is the degree of precision with which the position of the scanning mirror is known. The position of the mirror can be determined with a high degree of accuracy by employing a helium-neon (He-Ne) laser as a point of reference (Fig. 4.8).

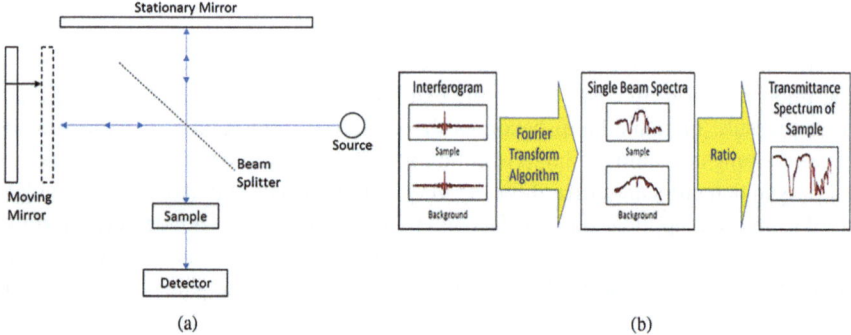

Fig. 4.8 (**a**) Internal working of FTIR showing a schematic of Michelson interferometer. (**b**) FTIR Spectroscopy principle and transformation of the spectrum. (Siddiqui et al. 2023)

4.6.7 Computers and Spectra

In current IR instruments, the computer is an essential component that serves a variety of purposes and is necessary for their operation. The computer is responsible for controlling the instrument; for instance, it determines the scan speeds and spectral limits, as well as the scanning process (start and stop). The spectrum is digitized as a result of the fact that it reads spectra from the instrument into the memory of the computer while the spectrum is being scanned. The computer can be used to alter spectra in a variety of ways, such as by adding and deleting spectra or by expanding sections of the spectrum that are of interest. Continuously scanning the spectra and either averaging or adding the results to the computer's memory are also functions that are performed by the computer. Complex analyses can be executed automatically by following a set of pre-programmed commands. The earliest IR instruments were able to record the percentage of transmittance throughout a linear range of wavelengths. At this point, it is not normal practice to use wavelength for everyday samples; instead, the wavenumber scale is frequently utilized. The spectrum is the name given to the output that is produced by the instrument. A spectrum is presented by the majority of commercial instruments, with the wavenumber higher to lower from left to right through the spectrum. Three primary regions may be distinguished within the IR spectrum (near-IR, mid-IR, and far-IR). Figure 4.9 shows the regions of the IR spectrum.

Despite the fact that the mid-IR area is utilized in a multitude of IR applications, the near-IR and far-IR regions simultaneously offer valuable information regarding particular substances. In general, the region between 4000 and 1600 cm^{-1} has a lesser number of IR bands (known as functional group region), while regions between 1600 and 400 cm^{-1} have a greater number of bands (known as fingerprint

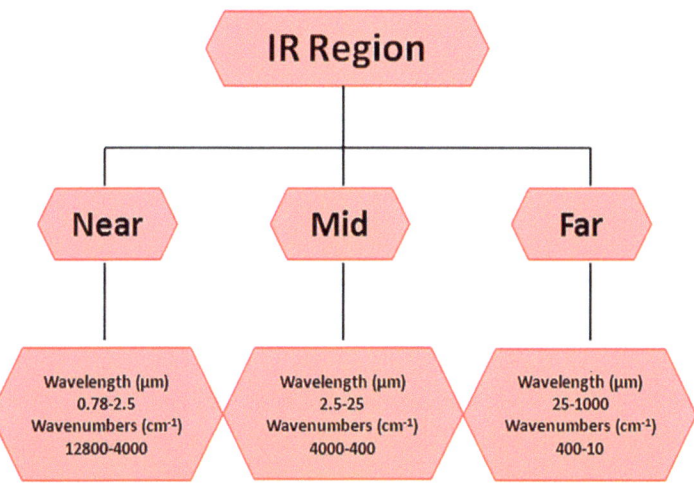

Fig. 4.9 The full range of the infrared region (mid, near, and far)

region). There are instances in which the scale is altered in such a way that the region between 4000 and 1800 cm^{-1} is reduced in size, while the region between 1800 and 400 cm^{-1} is enlarged in order to accentuate features that are of importance. The vertical scale can be displayed as a percentage of transmittance, with 100% representing the highest point on the spectrum. Having the option to choose between absorbance and transmittance as a means of measuring band intensity is a typical practice. The IR spectra of edible oil (soybean) demonstrate the change in appearance between absorbance and transmittance spectra, respectively. Figure 4.10 illustrates the difference in appearance between the transmittance and absorbance spectra of soybean oil in the mid-IR range, whereas Table 4.3 shows the typical (17) functional groups identified in the oils and fats. The choice between the two modes ultimately depends on personal preference. However, transmittance is commonly employed for spectral interpretation (qualitative), whereas absorbance is utilized for quantitative analysis.

4.6.8 Transmission Methods

Transmission spectroscopy is the IR approach that has been around the longest and is the easiest to understand. A sample absorbs IR light of a certain wavelength as it travels through the sample, which is the foundation of this technique. Through the utilization of this method, it is feasible to conduct analyses on samples that are in the liquid, solid, or gaseous states. There are a number of distinct varieties of

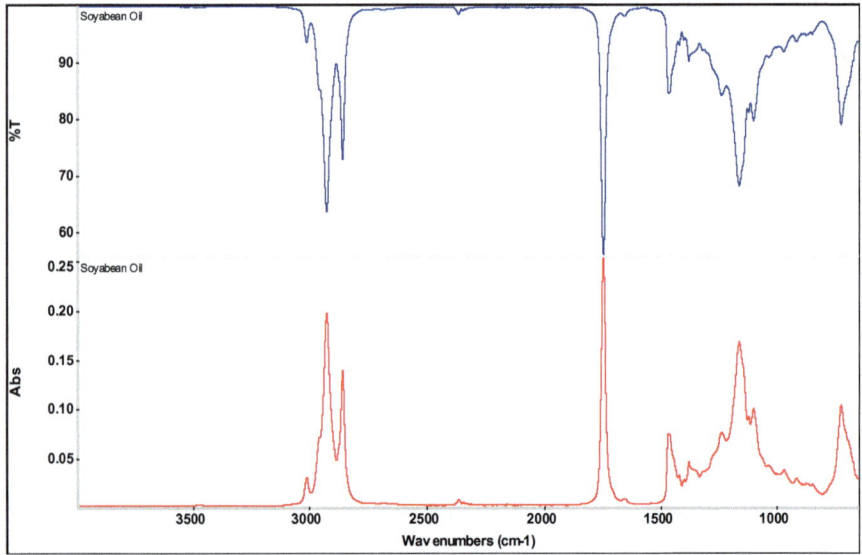

Fig. 4.10 Absorbance and transmittance mid-IR spectrum of soybean oil

Table 4.3 Typical functional groups identified in oils and fats

S. No.	Range (cm^{-1})	Functional group	Nominal frequency	Mode of vibration
1	3029–2989	=C–H (cis)	3010	Stretching
2	2989–2946	–C–H (CH3)	2954	Stretching (asym)
3	2946–2881	–C–H (CH2)	2921	Stretching (asym)
4	2881–2782	–C–H (CH2)	2852	Stretching (sym)
5	1795–1677	–C=O (ester) –C=O (acid)	1743 1710	Stretching Stretching
6	1670–1620	–C=C (cis)	1655	Stretching
7	1486–1446	–C–H (CH2)	1465	Bending (scissoring)
8	1446–1425	–C–H (CH3)	1456	Bending (asym)
9	1425–1409	=C–H (cis)	1416	Bending (rocking)
10	1382–1371	–C–H (CH3)	1377[a]	Bending (sym)
11	1290–1211	–C–O –CH2–	1238[a]	Stretching Bending
12	1211–1147	–C–O –CH2–	1160[a]	Stretching Bending
13	1128–1106	–C–O	1117[a]	Stretching
14	1106–1072	–C–O	1098[a]	Stretching
15	1006–929	–HC=CH– (*trans*)	967[a]	Bending (out of plane)
16	929–885	–HC=CH– (cis)	916[b]	Bending (out of plane)
17	754–701	–(CH2)n– –HC=CH– (cis)	721[a]	Rocking bending (out of plane)

[a]Guillén and Cabo (1999) and [b]Silverstein et al. (1981)

transmission solution cells that may be purchased. Despite the fact that they are beneficial for volatile liquids, fixed path-length sealed cells cannot be disassembled for inspection or cleaning. In order to facilitate the cleaning of the windows, semi-permanent cells can be demounted after use. The demountable cell contains front and back plates, a gasket, front and back windows, and a spacer. Because the spacer usually consists of polytetrafluoroethylene (PTFE, sometimes known as "Teflon") and is available in a variety of thicknesses, it is possible to use a single cell for a number of different path lengths. Variable path length cells are equipped with a mechanism that allows for continuous changes of the path length, whereas a vernier scale enables precise adjustment. A syringe is used to fill each of these different types of cells, and before beginning the sampling process, the ports of the syringe are sealed with PTFE plugs.

The type of window material is an important factor to take into account when selecting IR cells. The window material must be transparent to the IR radiation that is incident on it, and alkali halides are typically utilized in transmission methods. Sodium chloride (NaCl) is the most cost-effective material, but other materials that are commonly used are listed in Table 4.4. Within the realm of IR spectroscopy, the utilization of water as a solvent poses a number of challenges. The water's IR modes

Table 4.4 Summary of optical materials used in transmission infrared spectroscopy (Barbara Stuart 2004)

Window material (useful range)	Refractive Index	Solubility	Advantage/disadvantage
CaF_2 77,000–900 cm^{-1}	1.4	Insoluble in water; resist most acids and bases	Useful for high-pressure work/ doesn't fog
BaF_2 66,666–800 cm^{-1}	1.5	Insoluble in water; soluble in acids and ammonium chloride	Sensitive to thermal and mechanical shock/doesn't fog
KBr 43,500–400 cm^{-1}	1.5	Soluble in water and alcohol; slightly soluble in ether	Good resistance to thermal and mechanical shock/hygroscopic
NaCl 40,000–600 cm^{-1}	1.5	Soluble in water Slightly soluble in alcohol	Fair resistance to thermal and mechanical shock/low cost and easily polished
KCl 33,000–400 cm^{-1}	1.5	Less solubility in water and alcohol	Fair resistance to thermal and mechanical shock/low cost and easily polished Hygroscopic
CsBr 42,000–250 cm^{-1}	1.7	Soluble in water and acids	Preferred in far IR applications, harder, more resistant to scratches/ hygroscopic
CsI 42,000–200 cm^{-1}	1.7	Soluble in water and alcohol	Preferred in far IR applications, softer/hygroscopic

are extremely strong, and there is a possibility that they will overlap with the sample modes of interest. Given that NaCl is soluble in water, it is not possible to utilize it as a material for an IR window in situations where water is used as a solvent. Small path lengths, typically 10 micrometers, are accessible in liquid cells, which contribute to the reduction of the intensities of the extremely powerful IR modes that are generated in the water spectrum. Due to the fact that the path length is relatively short, the sample cavity is also relatively small, which enables the examination of samples in milligram quantities.

Examining liquid samples can be accomplished in a short amount of time using liquid films. One such configuration involves placing a drop of liquid in between two IR plates, which are then positioned inside a cell holder. Choosing an appropriate solvent is necessary before beginning the process of generating an IR sample in solution. When selecting a solvent for a sample, it is essential to take into consideration the following aspects: The solvent must be able to dissolve the compound, and it must be as non-polar as possible in order to reduce the number of interactions between the solute and the solvent, and it must not substantially absorb IR light. When conducting quantitative analysis on a sample, it is essential to make use of a cell that has a path length that is already defined.

Three general methods are utilized for the examination of solid samples when conducting transmission IR spectroscopy. These methods are alkali halide discs, mulls, and films. When selecting a method, the type of sample that is going to be

4.6 Spectroscopic Techniques for Food Authentication

analyzed is a very important factor to consider. Mixing a solid sample with a dry alkali halide powder [commonly used (KBr) potassium bromide] is the first step in the process of using alkali halide discs. Once the mixture is ground, it is typically placed in an evacuated die and subjected to a pressure of around 10 tons. This process produces a clear transparent disc, and KBr exhibits full transparency within the mid-IR region. When making alkali halide discs, it is imperative to take into account specific parameters.

- The ratio of the sample to alkali halide.
- Too thick disc (transmit too little radiation).
- Too thin disc (fragile and difficult to handle).
- Size of the crystals (sample or alkali halide).
- Moisture (sample or alkali halide).

Keeping in view the above parameters, a disk with a diameter of around 1 cm, formed by combining 2 to 3 mg of a sample with around 200 mg of halide, often yields an adequate thickness of approximately 1 mm.

The mull method for solid samples entails the processing of the sample through grinding, followed by a suspending of approximately 50 mg of the sample in 1 to 2 drops of a mulling agent. Subsequently, the process is continued until a homogeneous paste is achieved. Nujol, sometimes known as liquid paraffin, is often employed as the predominant mulling agent. While the mull approach is efficient and straightforward, it is important to take into account certain experimental factors.

- The ratio of the sample to mulling agent.
- Too little sample (weak spectrum with strongest absorption bands only).
- Too much sample (poor transmission of radiation).

One primary drawback associated with this approach is the potential obfuscation of bands that may be inherent in the substance under analysis due to the presence of mineral oil.

Films can be generated through two primary methods: solvent casting and melt casting. Solvent casting involves dissolving the sample in a suitable solvent, with the concentration determined by the desired film thickness. It is necessary to select a solvent that can dissolve the sample and create a consistent film. The solution is carefully poured onto a flat metal plate or a glass plate and evenly distributed to achieve a consistent thickness. Subsequently, the solvent can be evaporated in an oven, and after achieving dryness, the film can be removed from the plate. It is important to use caution when exposing samples to heat, as it can potentially lead to degradation. In an alternative approach, it is feasible to apply a film directly onto the IR window for direct utilization. Melt casting can be used to create solid samples that melt at comparatively low temperatures without breaking down. The preparation of a film involves the application of heat to the sample within a hydraulic press, which consists of heated metal plates (Barbara Stuart 2004).

4.6.9 Reflectance Methods

When traditional transmittance methods fail to provide useful results for analyzing a sample, reflectance techniques may be employed. There are two distinct categories in which reflectance methods can be classified. An attenuated total reflectance cell can be used to detect internal reflectance by placing it in contact with the sample. Additionally, a range of external reflectance assays exist that entail the direct reflection of an IR beam from the surface of the sample.

4.6.10 Attenuated Total Reflectance Spectroscopy

The phenomenon of total internal reflection is utilized in attenuated total reflectance (ATR) spectroscopy. Total internal reflection occurs when the angle of incidence at the interface between the sample and the crystal is greater than the critical angle (Du and Zhou 2009). The critical angle is a function of the refractive indices of the two surfaces, and it is the angle at which the beam of radiation enters the crystal that determines whether or not it will experience total internal reflection. The spectrometer measures and plots the attenuated light as a function of wavelength, which determines the absorption spectral properties of the material. When a material that selectively absorbs radiation comes in close contact with a reflecting surface, the beam loses energy at the wavelength where the substance absorbs as it penetrates a fraction of a wavelength beyond the reflecting surface.

ATR cell crystals are composed of substances that are exceedingly aqueous, insoluble, and have an extremely high refractive index. The aforementioned materials encompass zinc sulphide (ZnS), thallium-iodide (KRS-5), zinc selenide (ZnSe), diamond, AMTIR (GeAsSe), silicon (Si), and germanium (Ge). Table 4.5 provides a summary of the features exhibited by the materials widely employed in the ATR crystals. Various configurations of ATR cells enable the analysis of both liquid and solid materials. By incorporating an inlet and an outlet into the device, a flow-through ATR cell can also be built. Various configurations of ATR cells enable the analysis of both liquid and solid materials. By incorporating an inlet and an outlet into the device, a flow-through ATR cell can also be built. Consequently, this makes it possible for solutions to flow continuously through the cell and enables spectral changes to be tracked over time. Multiple internal reflectance (MIR) is a technology that is comparable to ATR; however, MIR generates more intense spectra from multiple reflections due to its implementation. A prism is normally utilized for ATR work; however, for MIR work, specially shaped crystals are used. These crystals create a large number of internal reflections >25.

Table 4.5 Summary of crystal materials used in ATR-IR spectroscopy

Crystals (useful range)	Refractive index	Depth of penetration	Angle of incidence	Hardness (Kg/mm^2)	pH range	Physical characteristics	Solubility
ZnS	2.2	3.86	45°, 60°	240	5–9	Withstand thermal and mechanical shock	Insoluble in the water, except KRS-5 (slightly soluble)
KRS-5 (TlBrI) 18,800–200 cm^{-1}	2.37	2.13	45°	40	5–8	–	
ZnSe 20,000–650 cm^{-1}	2.4	2.0	40°, 45°, 50°, 55°, 60°, 70°	120	5–9	Hard, brittle	
Diamond	2.4	2.0	30°, 45°	5700	1–14	Hard	
AMTIR (GeAsSe) 14,285–833 cm^{-1}	2.5	1.70	45°	170	1–9	Relatively hard, brittle	
Silicon 8300–660 cm^{-1}	3.4	0.85	45°	1150	1–12	Hard and brittle, withstand thermal shock, inert	
Ge 5500–675 cm^{-1}	4.0	0.66	30°, 40°, 45°, 60°	550	1–14	Hard and brittle, sensitive to temperature, reflection losses	

Source: https://www.piketech.com/atr-crystal-selection/,https://jascoinc.com/learning-center/theory/spectroscopy-1/attenuated-total-reflectance/

4.6.11 Applications of Transmission and ATR Spectroscopy

For the analysis of thin films, gases, and solids where the IR light can pass through the sample, transmission spectroscopy is ideal. It is widely utilized in material characterization, environmental monitoring, and organic and inorganic chemistry. For quantitative analysis and impurity detection in transparent samples, this technique is particularly effective. However, without requiring a lot of sample preparation, ATR spectroscopy is widely used to analyze solids, liquids, powders, and pastes. It is frequently used in food quality testing, pharmaceuticals, polymers, and forensics. ATR is also useful for studying surface modifications, coatings, and biological materials like skin or tissues. The benefits and drawbacks of both analytical techniques are displayed in Fig. 4.11.

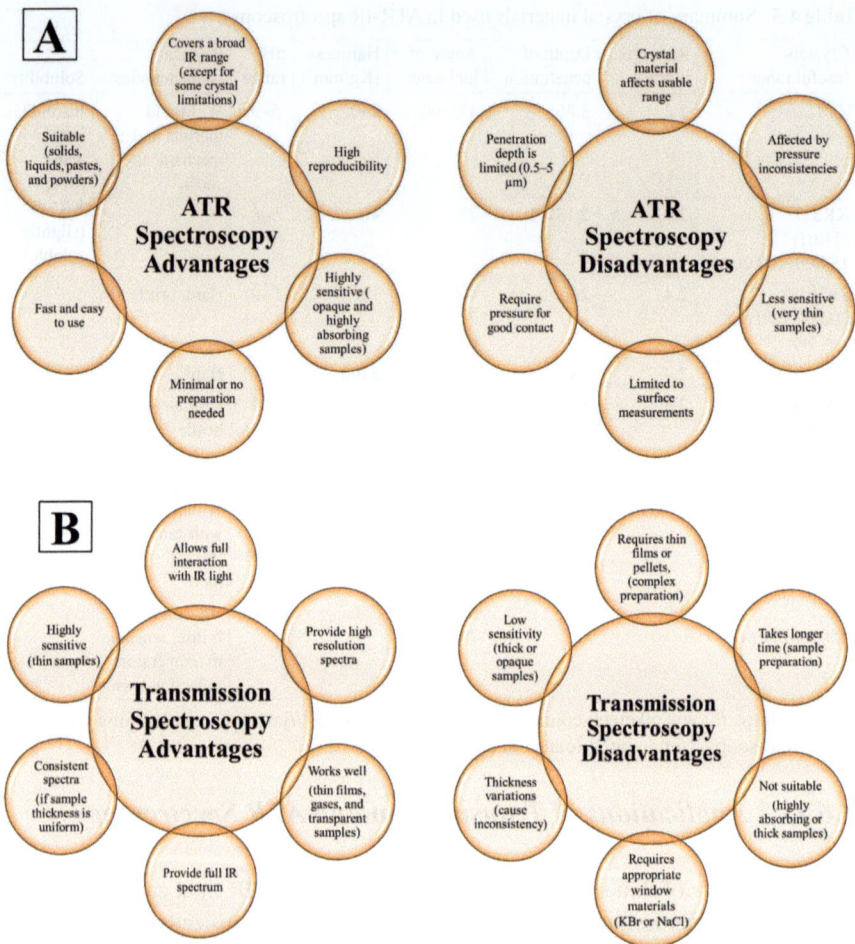

Fig. 4.11 Advantages and disadvantages of ATR spectroscopy (**a**) and Transmission Spectroscopy (**b**)

4.6.12 FTIR Spectroscopy and Chemometrics

Among molecular spectroscopies, FTIR spectroscopy is the most reported method used for the analysis of non-Halal components in any product due to its being rapid, nondestructive, sensitive, and easy to prepare the sample. FTIR spectroscopy, coupled with chemometrics, is extensively utilized for the swift identification of non-Halal components because of its benefits as a fingerprint analytical approach (Rohman and Windarsih 2020). The FTIR spectra acquired during authentication examination are intricate and challenging to interpret; however, advancements in chemometrics software and computational technology can help overcome these challenges. The variables employed in the development of multivariate models for

4.6 Spectroscopic Techniques for Food Authentication

the authentication study of fats and oils are absorbance values within the designated region of FTIR spectra (Rohman et al. 2016). The adulteration of oils and fats is challenging to detect visually, as the spectra of adulterated products closely resemble those of real oils and fats. Nonetheless, variations in absorbance values are necessary in specific regions due to the differing composition of adulterated samples. The incorporation of different oils and fats significantly alters the chemical compositions in adulterated samples. Certain substances may be present at elevated concentrations, whilst others may appear in diminished concentrations (Hassan et al. 2018; Siska et al. 2023). Selected FTIR fingerprints utilized in multivariate models are associated with variations in specific compounds within the samples. As a result, differences in absorbance values or shift in wavenumber vibrations can be observed. In the authentication of fats and oils, different chemometric techniques are frequently applied to interpret complex chemical data, as illustrated in Fig. 4.12.

Various data pre-processing techniques, such as mean centering, Savitzky–Golay derivatization, standard normal variate, baseline corrections, signal correction and compression, spectra normalizations, and multiplicative correction, are applied to achieve optimal analytical results (Rohman and Windarsih 2020). Chemometrics approaches used in food authentication are categorized into two types: pattern recognition methods (e.g., PCA, CA, DA) and multivariate calibration techniques (e.g., PLS, PCR, etc.). The combination of FTIR spectroscopy and chemometrics provides an effective approach for analyzing fats, including those extracted from meat in food products. To detect non-Halal meats in processed foods, lipids are extracted using suitable solvents and techniques before undergoing instrumental analysis. Meat-containing food products are hydrolyzed, and their fat content is extracted for

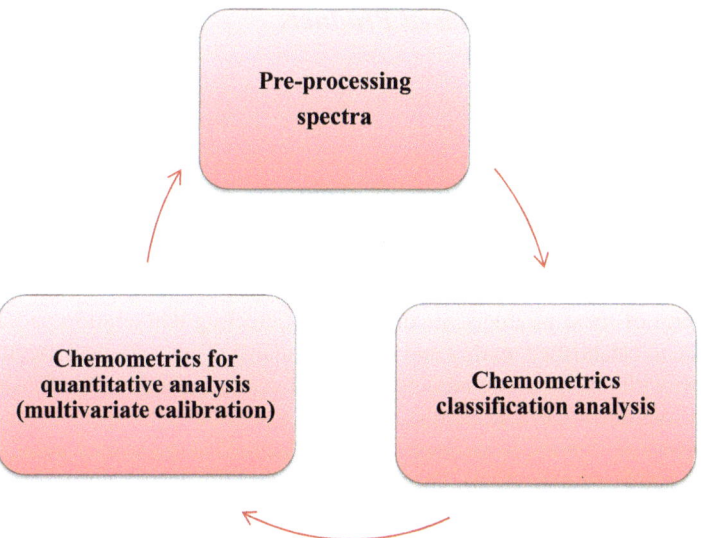

Fig. 4.12 Illustrating the various chemometric techniques are commonly used to analyze complex chemical data

FTIR analysis. Figure 4.3 presents a schematic representation of the analytical procedure for analyzing meat-based food products using FTIR spectroscopy integrated with the chemometric method.

4.7 FTIR Spectroscopy for Halal Authentication

FTIR spectroscopy is a crucial tool for Halal authentication, offering rapid, non-destructive detection of non-Halal adulterants across various products. In the meat industry, FTIR identifies non-Halal fats like lard in meatballs and processed meats, distinguishing pork-derived fats from Halal alternatives such as beef tallow and chicken fat (Ahmad and Ayub 2022). It also authenticates gelatin sources, differentiating porcine from bovine origins, and verifies confectionery items like chocolates, candies, and marshmallows containing gelatin, emulsifiers, and flavorings. FTIR effectively detects lard or pork adulteration in edible oils and identifies hidden animal-derived ingredients or alcohol-based compounds in cosmetics and personal care products. In beverages, including fruit juices, soft drinks, and alcoholic drinks, FTIR quantifies alcohol content by targeting key absorption bands, ensuring compliance with Halal standards (Fig. 4.13). With the support of chemometric methods like PLSR and PCA, FTIR provides a reliable, real-time approach to Halal authentication. This technology enhances food and product integrity, supports regulatory compliance, and strengthens consumer confidence (Rohman 2018).

4.7.1 Meat and Meat-Based Products

In the global food sector, the authenticity of meat and meat products is a major problem. Reliable techniques are required to ensure the integrity of meat products due to issues including economic adulteration, ethical considerations, religious conformity, and food safety. Vibrational spectroscopy, encompassing techniques like FTIR spectroscopy, has emerged as a promising tool for the authentication of meat and meat products (Hossaina et al. 2021; Rohman 2018). Several studies have explored the use of FTIR spectroscopy as a reliable technique for verifying the authenticity of meat products, particularly in detecting adulteration and ensuring adherence to established quality and safety standards. This analytical method, often combined with chemometric modeling, has proven to be highly effective in identifying various forms of adulterants, including non-Halal substances and economically motivated contaminants such as pork and lard in processed meat products (Rohman and Windarsih 2020). The integration of FTIR spectroscopy with advanced statistical approaches enhances its precision, making it a valuable tool for regulatory bodies and food industries in maintaining product integrity and consumer trust (Table 4.6).

4.7 FTIR Spectroscopy for Halal Authentication

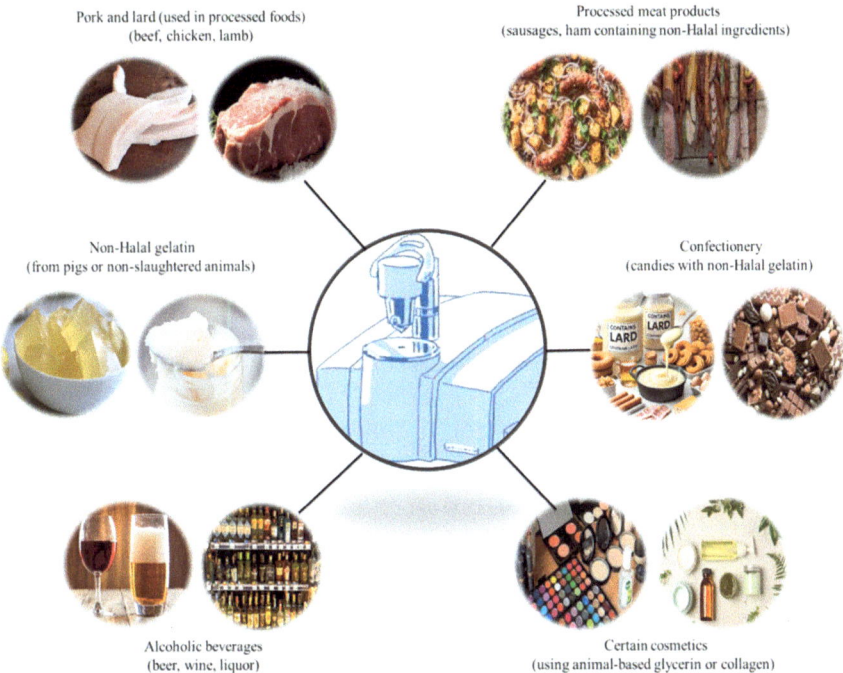

Fig. 4.13 Applications of FTIR spectroscopy in the authentication of Halal and non-Halal products

Al-Jowder et al. (1997) employed FTIR-ATR with PCA for the authentication of meat (chicken, turkey, and pork) as well as the detection of frozen-thawed meat. PCA was applied to IR truncated to 448 data points within the 1000 to 1800 cm^{-1} region, successfully differentiating between minced chicken, turkey, and pork. Similarly, Rannou and Downey (1997) conducted a spectroscopic study to classify chicken meat, turkey, and pork using visible, NIR, and MIR spectroscopy. A total of 74 meat samples were analyzed, and discriminant models were developed separately for MIR and visible-NIR spectra, with specific spectral regions investigated for optimal classification. The best predictive model using MIR spectra achieved an 86.5% accuracy (32 out of 37 test samples), while the visible-NIR model attained a higher classification accuracy of 91.9% (34 out of 37 samples) within the 400–1100 nm wavelength range. In a later study by Al-Jowder et al. (1999), FTIR successfully detected the adulteration of raw minced beef with specific offal types from the same species, specifically ox kidney and liver. PCA analysis clearly distinguished between muscle, kidney, and liver, and predictive models were created using the combined methods of PLS and CVA to differentiate the specimens of beef, kidney, and liver within the wavenumber 900–1800 cm^{-1}. Likewise, Ding and Xu (2000) developed a method using NIR spectroscopy to detect adulteration in beef hamburgers with pork, mutton, skim milk powder, or wheat flour, achieving an accuracy of up to 92.7%. The detection accuracy improved as the level of

Table 4.6 Application of FTIR spectroscopy combined with chemometrics for authentication analysis of meat-based products

Adulterated meat/food products	Meat adulterants	Spectral range	Spectral treatment and chemometrics	References
Chicken/Turkey meat	Pork	1000–1800 cm^{-1}	PCA	Al-Jowder et al. (1997)
Chicken/Turkey meat	Pork	400–1100 nm	PCA	Rannou and Downey (1997)
Beef hamburgers	Pork	–	–	Ding and Xu (2000)
Beef	Pork	1800 cm^{-1} and 1600 cm^{-1}	–	Yang and Irudayaraj (2001a, b)
Ground beef	Bovine spinal cord	400–4000 cm^{-1}	PLS	Gangidi et al. (2003)
Raw meat	Chicken, Turkey meat, beef, pork, and lamb	400–2500 nm	SIMCA	Arnalds et al. (2004)
Beef muscle species	Pork	400–2500 nm	PCA, PLS	Cozzolino and Murray (2004)
Beef/chicken/Turkey meat	Pork/lamb	942, 988, and 1606 cm^{-1}	PCA	Ellis et al. (2005)
Chicken meat	Pork	12,500 to 4000 cm^{-1}	PLS/PCA	Fan et al. (2010)
Beef meat	Pork	1100–2500 nm	PLS/PCA/SLDA	Restaino et al. (2011)
Buffalo, sheep, and camel meat	Pork	4000–400 cm^{-1}	–	Lamyaa (2013)
Beef	Pork	900–1900 cm^{-1}	PLS	Abu-Ghoush et al. (2017)
Beef meat	Pork	1002 to 1240, 1700 to 1714, 1764 to 1795 (BO), and 1105 to 1182 (PO)	PCA, PLS	Hu et al. (2017)
Ground meat	Beef, lamb, and pork	1720 and 1770 nm	SIMCA, OCPLS, SNV, MSC	Pieszczek et al. (2018)
Beef mutton	Pork	2925, 1464, and 1173 cm^{-1}	PLS-DA	Yang et al. (2018)
Minced beef	Pork	900–1700 nm and 400–1000 nm	PLS-DA, FFNN, SVM	Rady and Adedeji (2018)

(continued)

4.7 FTIR Spectroscopy for Halal Authentication

Table 4.6 (continued)

Adulterated meat/food products	Meat adulterants	Spectral range	Spectral treatment and chemometrics	References
Minced beef	Duck meat and pork	12,500–5400 cm^{-1}	DA and PLS	Leng et al. (2020)
Beef, mutton, lamb, camel, chicken, and veal	Pork	10,000 to 4000 cm^{-1}	PLS-DA, PCA	Mabood et al. (2020)
Beef corned	Corned pork	1180–730 cm^{-1}	PLS, PCA	Guntarti et al. (2020)
Minced meat/offal/chicken	Pork/bovine	405–970 nm	MSI, SVM	Fengou et al. (2021)
Meat fat	Pork	–	DD-SIMCA/PCA	Totaro et al. (2023)
Beef and chicken fat	Pork	1745, 1116, 1550, and 722 cm^{-1}	–	Yulirohyami et al. (2023)
Beef, lamb, and chicken	Lard	800–3500 cm^{-1}	PCA, PLS	Siddiqui et al. (2023)

adulteration increased. Once an adulterant was identified, its concentration was estimated using calibration equations. The standard errors of cross-validation for predicting adulteration levels were 3.33% for mutton, 2.99% for pork, 0.92% for skim milk powder, and 0.57% for wheat flour, with corresponding coefficients of variance of 0.87, 0.89, 0.99, and 1.00, respectively. In contrast, Yang and Irudayaraj (2001a, b) investigated the potential of FTIR-PAS (Photoacoustic Spectroscopy) for characterizing beef and pork by examining functional groups at different depths within meat samples. The study compared FTIR-PAS with the conventional ATR method and found that PAS spectra were better resolved and more sensitive to moisture variations with wave numbers 1800 cm^{-1} and 1600 cm^{-1}.

In another study, Gangidi et al. (2003) employed ATR-FTIR spectroscopy and PLS to detect and quantify bovine spinal cord in ground beef. Mid-IR spectra of the samples were collected containing spinal cord concentrations ranging from 0 to 100 ppm. PLS models were developed using the 400–4000 cm^{-1} spectral range, achieving a strong calibration performance (R_{cal} = 0.94, RMSE = 11.24) and a validation accuracy of R_{val} = 0.87 (RMSE = 17.22). Later on, Arnalds et al. (2004) investigated the use of visible and NIR spectroscopy combined with hierarchical discriminant analysis to identify species in raw meat. The study aimed to classify five meat types (chicken, turkey meat, beef, pork, and lamb) using chemometrics techniques. Spectral data in the range of 400–2500 nm were collected and processed to optimize classification performance. In order to identify and verify various beef muscle species, Cozzolino and Murray (2004) employed visible and NIR spectroscopy. Muscle samples from pork ($n = 44$), beef ($n = 100$), chicken ($n = 48$), and lamb ($n = 140$) were homogenized and scanned using visible and NIR regions (400–2500 nm). To discriminate between various meat species, PLS and PCA models were generated. Over 80% of the meat sample muscles were properly categorized by the models based on their species, respectively. Another study by Ellis et al. (2005) employed both Raman and FTIR spectroscopy for the comparative analysis of various meat types, including pork, lamb (neck fillets), (boneless steaks), chicken (skinless breast fillets and legs with skin), beef (rump steaks), and turkey (legs with skin and skinless breast fillets). A genetic algorithm was utilized on the FTIR data to discover appropriate wave numbers for differentiating various meat varieties and varied cuts within a certain type. A clear differentiation between chicken breast, legs, and turkey was achieved at the three subsequent wave numbers: 942, 988, and 1606 cm^{-1}. In continuation of the earlier NIR studies, Fan et al. (2010) found that economic adulteration in pork has been detected using NIR spectroscopy ranging from 12,500 to 4000 cm^{-1}. Liver and chicken were added to pork samples in 10% increments. Pretreatment spectra and raw data were used, and predictions and quantitative analysis were performed. PLS regression with SNV pretreatment produced the best prediction result for pork adulterated with liver samples; the correlation coefficient (R value), RMSEC, and RMSEP were 0.97706, 0.0673, and 0.0732, respectively. PLS, using the raw spectra, produced the most effective model for pork meat adulterated with chicken samples; the RMSEP, RMSEC, and correlation coefficient (R value) were 0.0525, 0.122, and 0.98614, respectively.

The discrimination of meat pates based on animal species using NIRS with chemometric techniques has been investigated by Restaino et al. (2011). The study analyzed 100% beef, 100% pork, and binary mixtures (beef-pork, w/w) using a NIR (1100–2500 nm) in reflectance mode. Stepwise Linear Discriminant Analysis (SLDA) and PCA were applied for classification. Both pork pate and beef samples were correctly classified (100%), while binary mixtures achieved 72% classification accuracy, respectively. The differentiation of pork content in mixes containing raw minced buffalo and camel meat was observed by Lamyaa (2013). Using FTIR spectroscopy, the study examined the structural characteristics of separate meats (camel, buffalo, sheep, and pork) as well as various mixes of pork in buffalo and pork in camel (10%, 30%, 50%, and 70% w/w) in the 4000–400 cm^{-1} spectrum region. The findings indicated that the highest protein level was found in camel meat, which declined in the order of buffalo > sheep > pork. The relative protein-to-lipid content of the pork-in-camel and pork-in-buffalo combinations dropped as the amount of pork in them increased. Similarly, Abu-Ghoush et al. (2017) presented a novel application of PLS-Kernel with MIR spectroscopy for rapid pork detection in beef mixtures down to 1.4 wt%. Analysis of raw spectra revealed a nonlinear correlation between pork concentration and IR band intensities of proteins and fats, particularly within the 900–1900 cm^{-1} spectral range. Chemometric modeling demonstrated the superior performance of PLS-Kernel over PCR and traditional PLS, requiring fewer computational resources while maintaining high accuracy. On the other hand, Hu et al. (2017) developed an optimal procedure for the investigation of ground beef meat adulterated with different types of beef and pork offal using FT-IR spectroscopy. Chemometric models for classification and quantification were applied within the wave number ranges 1002 to 1240, 1700 to 1714, 1764 to 1795 (beef offal), and 1105 to 1182 (pork offal). FT-IR spectroscopy, combined with a chemometric model, effectively quantified five different types of offal with an R^2 value exceeding 0.81. It distinguished pure beef from adulterated samples with an accuracy of over 99% and identified the specific type of offal present with more than 80% confidence.

In 2018, a study reported by Pieszczek et al. investigated and identified ground meat species for classifying (beef, lamb, and pork) using NIRS combined with class modeling techniques with a range between 1720 and 1770 nm. For improved classification accuracy, various models were developed, such as SNV, One-Class Partial Least Squares (OCPLS), Soft Independent Modeling of Class Analogy (SIMCA), Multiplicative Scatter Correction (MSC), and Inverse Scatter Correction (ISC). The results obtained validated that the NIR spectroscopic fingerprints are effective for identifying species of ground meat. In the same year, Yang et al. (2018) developed an effective method for detecting pork adulteration in beef and mutton using IR spectroscopy combined with advanced chemometric models Partial Least Squares Discriminant Analysis (PLS-DA) and Support Vector Machine (SVM) within the wave numbers 2925, 1464, and 1173 cm^{-1} (Fig. 4.14a, b). The PLS-DA model outperformed the SVM, achieving a coefficient of determination (R^2) of 0.99, with 100% accuracy in distinguishing between pork, beef, and mutton samples. These findings underscore the robustness and reliability of the PLS-DA approach over SVM for adulteration analysis (Fig. 4.15).

Rady and Adedeji (2018) evaluated NIR (900–1700 nm) and Vis-NIR Spectroscopy (400–1000 nm) regions for detecting adulterants in minced beef and pork. The study examined animal-based (pork, chicken) and plant-based (wheat gluten, texturized vegetable protein) adulterants using machine learning techniques for classification and prediction. The optimal models using selected wavelengths improved classification accuracy, achieving 96–100% accuracy in detecting adulteration and 69–100% in identifying adulterant type. Predictive models for adulterant levels attained correlation coefficients ($r = 0.78$–0.86) with a ratio of performance to deviation (RPD = 1.19–1.98). Later on, for the detection of pork and duck meat in minced beef, Leng et al. (2020) described a rapid and non-destructive method by NIR technology (12,500–5400 cm^{-1}). DA and PLS models were optimized by selecting suitable spectral wavelengths and employing various spectral pretreatments for the detection and prediction of adulteration. The DA model, utilizing the selected wavelength without any preprocessing approaches, attained optimal results, achieving classification rates of 100% for the binary system and 91.5% for the ternary system. The RMSEP for the binary and ternary samples was 7.27 and 9.27, respectively, and the correlation coefficient (Rp) was 95.80% and 95.69% for the best PLS models with full-wavelength for predicting adulterant levels. The finding of this study revealed that NIR technology can be employed for both binary and ternary minced beef adulteration. In contrast, detecting and quantifying pork meat in other meat samples (beef, mutton, lamb, camel, chicken, and veal) by combining multivariate analysis and NIR reflectance spectroscopy by Mabood et al. 2020. The pork was added in triplicate to 5952 combination samples made from 39 different

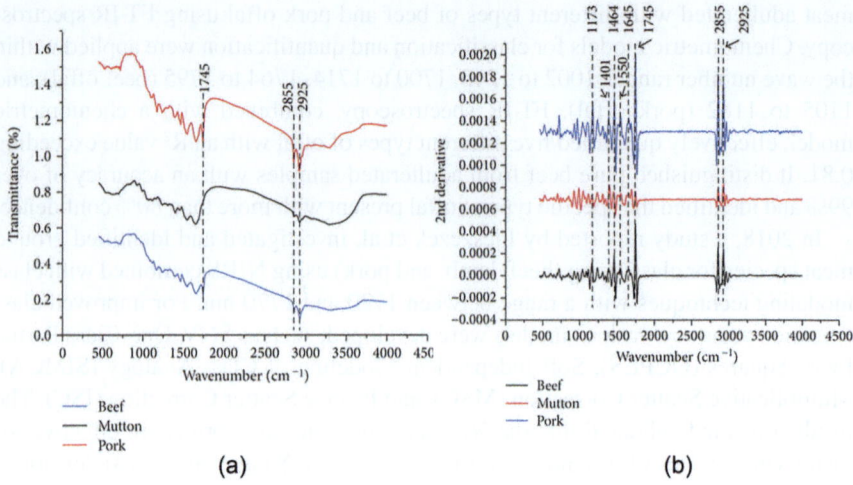

Fig. 4.14 (**a**) Average values of the absorbed spectrum, (**b**) 2nd derivative spectrum of pork, beef, and mutton. (Yang et al. 2018)

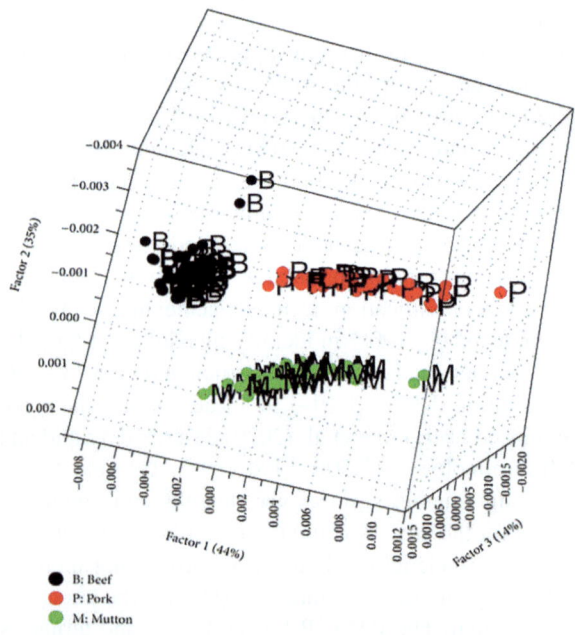

Fig. 4.15 Visualization of sample densities in 3 d data of the meat samples represented black, red, and green dots for B (beef), P (pork), and M (mutton), respectively. (Yang et al. 2018)

types of meat at 0%, 1%, 5%, 10%, 30%, 50%, 70%, 90%, and 100%. The wavenumber range in which the spectra were obtained was 10,000 to 4000 cm^{-1}. PLS-DA and PCA demonstrated strong discrimination between samples of pure and pork-spiked meat. The R^2 value was 0.9774, the RMSECV value was 1.08%, and the RMSEP value was 1.84%, respectively. Another study by Guntarti et al. (2020) utilized FTIR spectrophotometry with chemometrics to quantitatively examine a group of corned beef and corned pork. The corned pork beef reference samples come in seven different concentrations: 0%, 25%, 35%, 50%, 65%, 75%, and 100%. The selected regions for the examination of corned pork-beef fingerprints were 1180–730 cm^{-1}, with R^2 0.9833, RMSEC 2.06%, RMSEP 1.65%, and RMSECV 2.22%, according to the results of the quantitative analysis using chemometrics (PLS). The PCA findings indicated that 100% corned beef and corned pork were grouped in separate quadrants.

In a recent study, Fengou et al. (2021) evaluated spectroscopy-based sensors for detecting fraudulent substitutions in minced meat, specifically beef with bovine offal and pork with chicken (fresh and frozen-thawed samples) with wavelengths ranging from Vis (405 nm) to short-wave NIR (970 nm). Meat mixtures were prepared with 25% incremental adulteration, including pure samples, resulting in 120 total samples. SVM models were developed, with multi-spectral image (MSI) based models achieving the highest accuracy (87–100%), followed by Vis-based models (57–97%), while Fluo-based models showed the lowest performance in detecting adulteration. Meanwhile, Totaro et al. (2023) explored NIR spectroscopy for meat authentication based on rearing system, breed, and geographical origin. Fat samples from pork raised under extensive and intensive systems were analyzed using DD-SIMCA classification. SNV pre-treatment and four PCs yielded the best results, achieving 100% sensitivity and specificity in calibration and validation. The study highlighted the potential of NIRS as a rapid and reliable tool for meat quality authentication. Similarly, Yulirohyami et al. (2023) conducted a study to compare the fat profile (beef and chicken fat) of pork thighs using FT-IR. Based on the results, the IR pattern of the lard was identified based on the difference in absorption intensity. IR spectra established the existence of -C=C- bonds at 3005 and various other bands at 1745, 1116, 1550, and 722 cm^{-1}. On the other hand, Siddiqui et al. (2023) identified lard adulteration in beef, lamb, and chicken samples, with lipid extraction from pure and adulterated meats (10%–50% v/v) using FTIR spectroscopy (Fig. 4.16). PCA and PLS were applied to classify and develop a calibration model within the 800–3500 cm^{-1} range, respectively, by using four different regions of the spectrum, as shown in Figs. 4.17 and 4.18.

4.7.2 Meatball Food Products

Meatballs are among the most beloved and versatile food products, with a rich history that spans cultures and continents. Their origins are thought to date back to ancient times, with early versions appearing in various regions, each incorporating local ingredients and culinary traditions. The concept of shaping ground meat into balls for ease of cooking and consumption has made meatballs a universal dish, evolving to suit diverse palates and preferences. Adapted to suit local tastes, ingredients, and traditions, meatballs have become a globally cherished dish (Rianti et al. 2018; Augustyńska-Prejsnar et al. 2022). In Europe, meatballs became a hallmark of traditional cuisines. Swedish "köttbullar," typically made with a mix of beef and

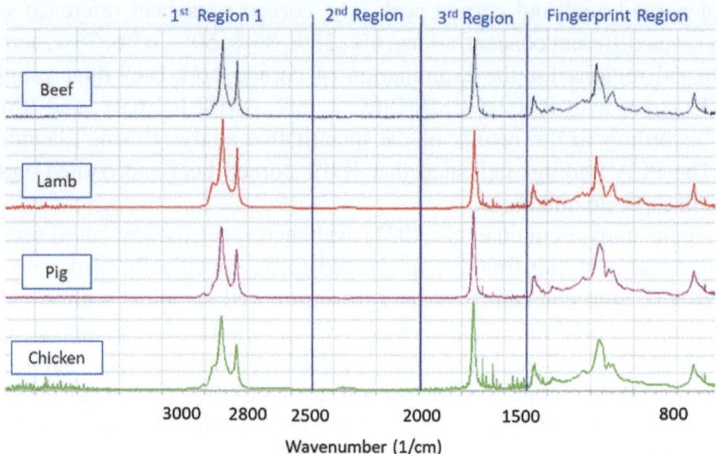

Fig. 4.16 FTIR spectra of pure beef, lamb, pig, and chicken in the range of 3500–650 cm^{-1} represent fingerprint and functional group areas. (Siddiqui et al. 2023)

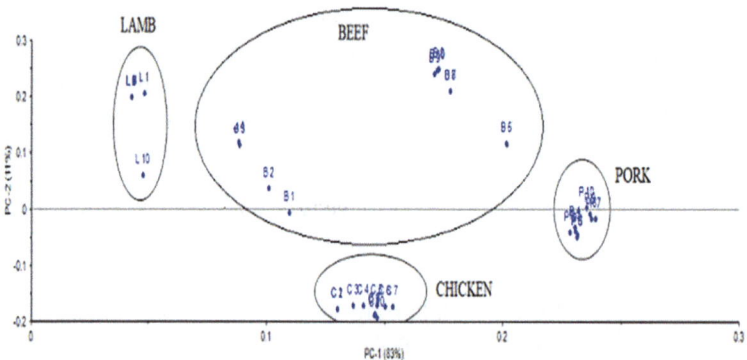

Fig. 4.17 Distribution of loadings values using PCA classification. (Siddiqui et al. 2023)

pork, served with gravy and lingonberry jam, gained global recognition. In Italy, "polpette" evolved as a family favorite, often paired with tomato-based sauces and pasta. In the Middle East and South Asia, "kofta" remains a popular dish, showcasing rich flavors from a blend of spices. In Persia, "kofta," a dish made from ground meat mixed with spices, was a staple in royal cuisine and eventually spread to neighboring regions. Similarly, the Roman cookbook *Apicius* described recipes for minced meat shaped into small balls, seasoned, and cooked in various ways (Huda et al. 2010; Guntarti et al. 2015). These traditions gave rise to regional adaptations as trade and migration introduced the concept to other parts of the world. From Sweden's *köttbullar* to Turkey's *köfte* and China's *Lion's Head*, meatballs are a versatile and universally enjoyed food (Table 4.7).

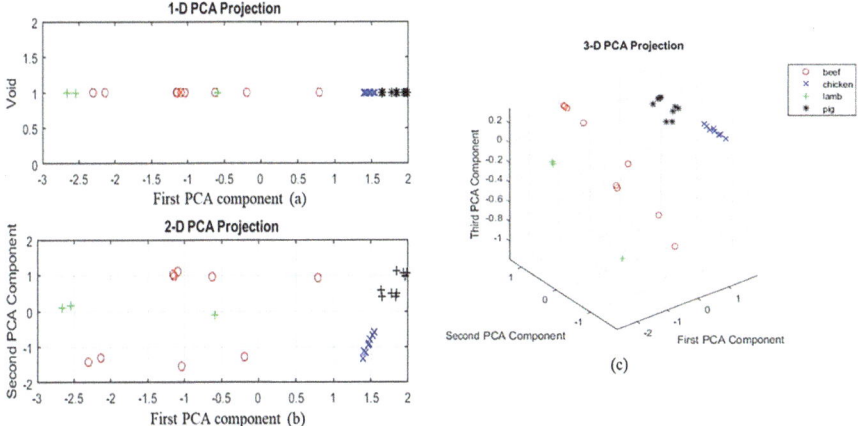

Fig. 4.18 (**a–c**) 3D projection of the first, second, and third PCs for all the samples for mid-IR regions. (Siddiqui et al. 2023)

Today, meatballs are a staple in both home cooking and commercial food production. They are made using a mixture of ground meat, typically beef, pork, chicken, or lamb, combined with seasonings, breadcrumbs, and sometimes eggs to bind the ingredients (Nakyinsige et al. 2012). The mixture is shaped into small balls and cooked through frying, baking, or simmering. Modern meatball recipes also include vegetarian and plant-based options, catering to evolving dietary preferences. Commercial meatball production has become a global industry driven by high consumer demand. Pre-packaged frozen or ready-to-eat meatballs are widely available, offering convenience to households and food service providers (Ahmad et al. 2024). However, the rise in mass production has also brought challenges, including concerns about meat adulteration and quality control, highlighting the need for robust food authentication practices.

The authentication of meat-based products, including meatballs, has become a crucial concern in the food industry. FTIR spectroscopy has proven to be a highly effective tool for detecting various adulterants in meatball products, including pork, lard, rat meat, wild boar meat, and dog meat (Fig. 4.19).

By targeting characteristic wavenumber regions and integrating advanced chemometric techniques such as PCA, this method allows for precise differentiation and quantification of adulterants. For instance, the fingerprint region (1500–600 cm^{-1}) has been extensively utilized to identify pork fat and lard due to their distinct spectral profiles, ensuring high sensitivity and accuracy in Halal food authentication (Rohman et al. 2011a, b). In the case of rat meat adulteration in beef meatballs, FTIR-ATR spectroscopy was employed, where fats were extracted using the Soxhlet method, evaporated, and analyzed using the ATR technique. PCA, applied to the wavenumber range of 1000–750 cm^{-1}, successfully distinguished between beef and rat meatballs (Rahmania and Rohman 2015). The same wavenumber range was also effective in detecting dog meat in meatball formulations by leveraging the molecular structure of fats and proteins (Rahayu et al. 2018a, b).

Table 4.7 Famous meatball dishes around the world

Continent	Country	Famous dish	Description and serving style
Asia	China	*Lion's Head Meatballs*	These are large pork meatballs served with bok choy in a flavorful broth
	India and Pakistan	*Kofta Curry*	In Indo-Pak, koftas are meatballs or vegetarian alternatives served in spiced curry gravy
	Middle eastern countries	*Kofta/Kufteh*	Ground lamb or beef meatballs mixed with herbs and spices. These are either grilled or cooked in a tomato-based sauce
	Vietnam	*Bò Viên*	Vietnamese beef meatballs are commonly served in pho (noodle soup) or grilled as a savory snack
	Indonesia and Malaysia	*Bakso*	A meatball dish made from beef or chicken usually served in a soup with noodles or rice and fried dumplings
Africa	Morocco	*Meatball Tagine*	Moroccan meatballs are cooked in a spiced tomato-based sauce and sometimes topped with eggs for extra richness
Europe	Albania	*Qofte*	Albanian meatballs are flavored with garlic, onions, and herbs. They can be served as snacks or part of a full meal
	Germany	*Königsberger Klopse*	German meatballs include capers and are served in a creamy white sauce with lemon for added flavor
	Greece	*Keftedes*	Greek-style meatballs are fried and seasoned with oregano and mint. They are often served with tzatziki or pita bread
	Italy	*Polpette*	Traditional Italian meatballs are made with beef or pork, breadcrumbs, herbs, and garlic
	Netherlands	*Bitterballen*	These are deep-fried meat snacks often served as appetizers with mustard for dipping.
	Spain	*Albóndigas*	Spanish meatballs are typically served in a rich tomato sauce or tapas, where they are presented in broth
	Sweden	*Köttbullar*	These are served with a creamy gravy, lingonberry sauce, and mashed potatoes
	Turkey	*Köfte*	Turkish köfte is made with ground lamb or beef and flavored with spices. They can be grilled, fried, or baked
America (South/North)	Brazil	*Almôndegas*	Brazilian meatballs are usually served alongside pasta or in a tomato-based sauce
	United States	*Spaghetti and Meatballs*	A well-loved Italian-American meal, where meatballs are paired with spaghetti and marinara sauce. Meatball sandwiches are also quite popular

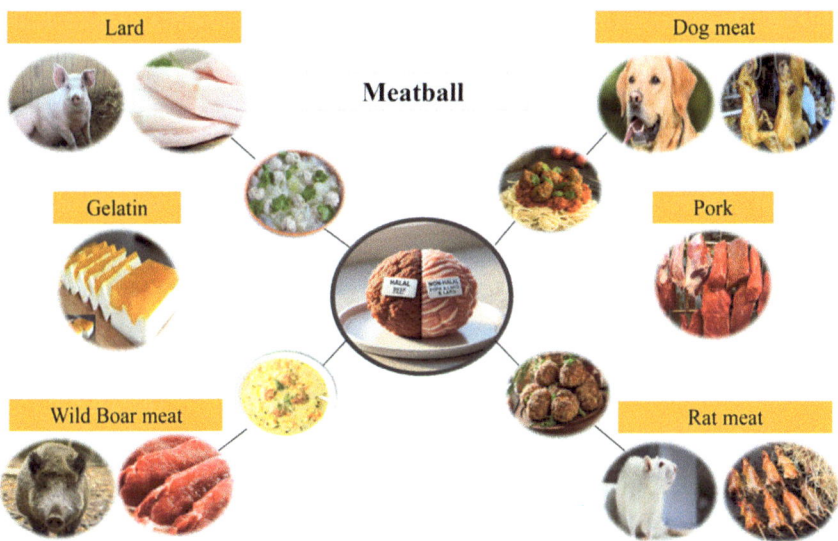

Fig. 4.19 Adulteration of meatballs with cheaper or non-Halal ingredients

Additionally, FTIR spectroscopy was applied to identify wild boar meat adulteration, with PCA using the wavenumber range from 1250–1000 cm^{-1}, achieving a clear classification between wild boar and beef meatballs (Ahda et al. 2020). These applications demonstrate the robustness and precision of FTIR spectroscopy in addressing meatball adulteration, ensuring food authenticity and compliance with Halal and quality standards. Numerous studies have investigated the application of FTIR spectroscopy for the authentication of meatball products, particularly in detecting adulteration and ensuring compliance with quality standards. FTIR spectroscopy, often integrated with advanced chemometric models, has demonstrated high efficacy in identifying non-Halal and economically motivated adulterants, including pork, lard, wild boar meat, and rat meat, within meatball formulations (Table 4.8).

Rohman et al. (2011a, b) investigated pork adulteration in beef meatballs using FTIR spectroscopy combined with PLS calibration (Fig. 4.20). The study demonstrated the effectiveness of FTIR in distinguishing pork fat from beef fat by analyzing spectral fingerprint regions (1200–1000 cm^{-1}), achieving a strong correlation (Fig. 4.21) between actual and predicted pork fat contents ($R^2 = 0.999$). Validation confirmed the method's reliability, with low prediction error (RMSE = 0.742), making it suitable for rapid and non-destructive detection of adulteration. This technique addresses significant religious (Halal and Kosher compliance) and economic concerns in food authenticity.

Table 4.8 The application of FTIR spectroscopy combined with chemometrics for authentication analysis of meatballs

Meat/food products	Adulterants	Spectral range	Spectral treatment and chemometrics	References
Beef meatballs	Pork	1200–1000 cm^{-1}	PLSR	Rohman et al. (2011a, b)
		1000–2500 nm		Vichasilp and Poungchompu 2014
		850–2000 nm	PLS and LDA	Kuswandi et al. (2015a, b)
		1200–1000 cm^{-1}	PCA and PLSR	Rohman et al. (2017)
		1022–883 cm^{-1}	PLS and PCA	Guntarti et al. (2018)
		1100–900 cm^{-1}		Nurani et al. (2022)
	Dog meat	1782–1623 and 1485–659 cm^{-1}	PLS	Rahayu et al. (2018a)
		1700–700 cm^{-1}	PLSR and PCA	Rahayu et al. (2018b)
	Rat meat	1000–750 cm^{-1}		Rahmania and Rohman (2015)
		750–1600 cm^{-1}		Guntarti and Prativi (2017)
		3100–800 cm^{-1}	PLS, LDA, and PCR	Lestari et al. (2022)
	Wild boar meat (WBM)	1250–1000 cm^{-1}	PLS and PCA	Guntarti et al. (2015)
		999–1481 cm^{-1} and 1650–1793 cm^{-1}		Ahda et al. (2020)
	Chicken fat, lard, and rat fat	3371, 3332, 2337, and 1743 cm^{-1}	–	Fajriati et al. (2021)
	Lard	1284–1018 cm^{-1}	PLS and PCA	Kurniawati et al. (2014)

Kurniawati et al. (2014) examined the detection and quantification of pork fat (lard) in meatball broth using FTIR spectroscopy combined with chemometric techniques, such as PLS calibration and PCA. The study identified wavenumber regions (1018–1284 cm^{-1}) that provided optimal accuracy for quantifying pork fat, achieving a coefficient of determination (R^2 = 0.9975) and low calibration error (RMSEC = 1.34%). PCA successfully classified pork and beef fat in broth, highlighting the ability to distinguish samples with high reliability. In another study, Vichasilp and Poungchompu (2014) explored the feasibility of detecting pork adulteration in meatballs (grilled and frozen) NIR spectroscopy (Figs. 4.22 and 4.23). The study utilized a wavelength range of 1000–2500 nm and developed PLS regression models to quantify adulteration. For beef meatballs, the model achieved an R^2-val of 0.88 with a standard error of cross-validation (SECV) of 3.45%, while for chicken meatballs, an R^2-val of 0.83 and SECV of 4.18% were observed. NIR acquisition at 25 °C for grilled samples provided reliable results, whereas frozen

4.7 FTIR Spectroscopy for Halal Authentication

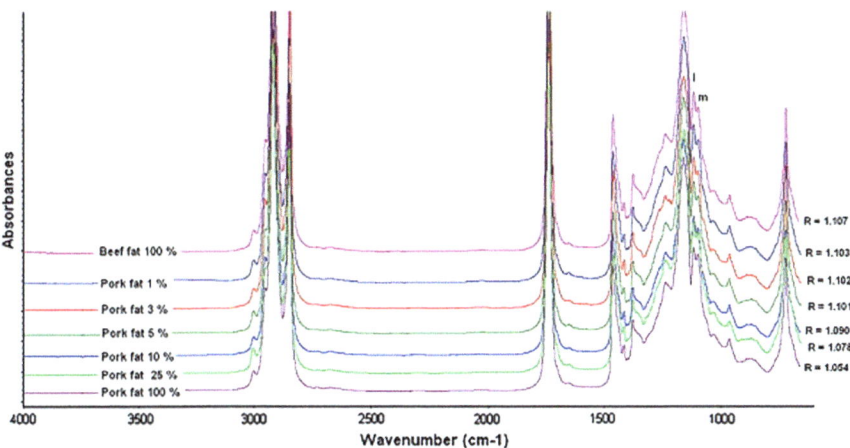

Fig. 4.20 FTIR spectra of beef and pork fats, as well as its mixture obtained from the extraction of meatball formulations, showing changes in the height ratio (R) at 1117 and 1097 cm^{-1} due to the increase of pork fat percentages at 4000–650 cm^{-1}. (Rohman et al. 2011a, b)

Fig. 4.21 Relationship between the actual value of PF and FTIR predicted value (% w/w) at selected fingerprint region (1200–1000 cm^{-1}). A = calibration; B = prediction. (Rohman et al. 2011a, b)

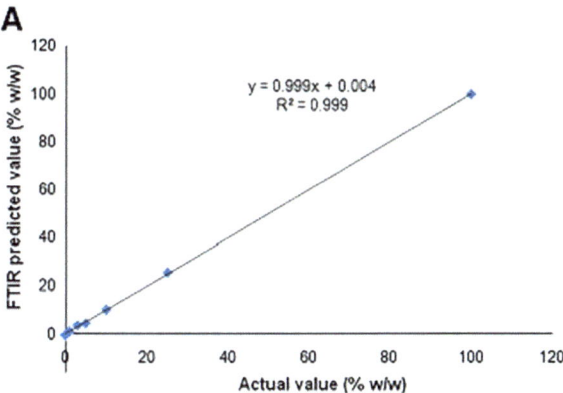

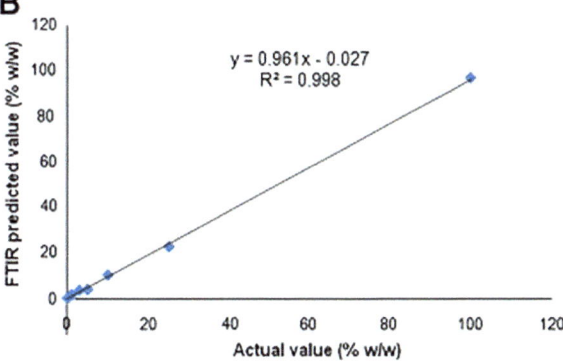

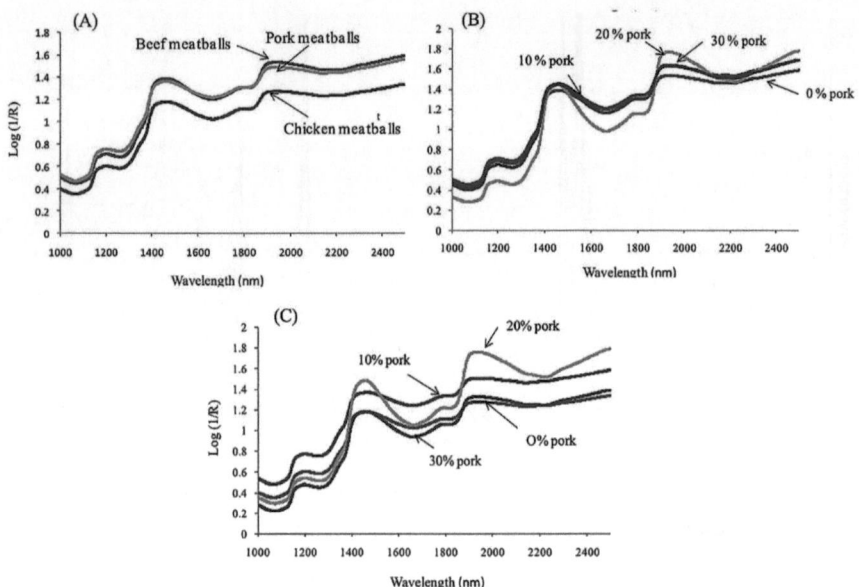

Fig. 4.22 The original spectra of beef, chicken and pork meatballs (**a**), the original spectra of beef meatballs with pork adulteration (**b**), and the original spectra of chicken meatballs with pork adulteration (**c**). (Vichasilp and Poungchompu 2014)

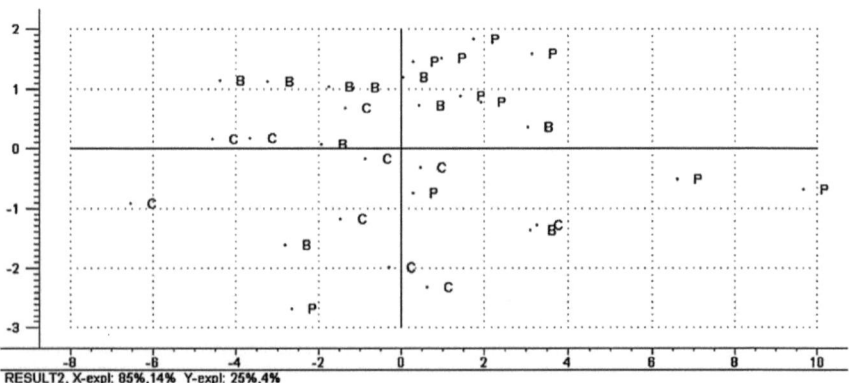

Fig. 4.23 The score plot of beef (B), chicken (C), and adulterated (P) meatballs. (Vichasilp and Poungchompu 2014)

samples yielded poor predictions. Despite relatively high errors, the study highlighted the advantages of NIR, including speed, low cost, and minimal sample preparation, making it a promising initial screening tool for detecting pork adulteration in Halal meat products.

Similarly, Kuswandi et al. (2015a, b) analyzed NIR spectra for detecting pork adulteration in beef meatballs using PLS and LDA models. Using the first derivative spectra, both models accurately classified 100% of the pork-adulterated beef meatball samples (training and test sets). In another study, Rohman et al. (2017) conducted a study to develop an FTIR spectrophotometric method combined with PCA and real-time PCR for determining pork-beef mixtures in meatballs. The results demonstrated that normal FTIR spectra at wavenumbers of 1200–1000 cm^{-1}, coupled with PLS and PCA, were effective for quantifying and classifying pork in beef meatballs. The relationship between the actual and predicted values of lard (the lipid fraction from meatballs containing pork) using the FTIR spectrophotometric method showed a strong correlation, with a coefficient of determination (R^2) of 0.997 and a standard error of calibration of 0.04%. Guntarti (2018) also utilized PLS and PCS for quantitative and classification analysis of pork and beef in meatball formulations (0, 25, 35, 50, 65, 75, and 100%). It was reported that the PLS analysis produced a linear regression equation, y = 0.9984x + 0.0758y = 0.9984x + 0.0758y = 0.9984x + 0.0758, with a determination coefficient (R^2) of 0.9984, (RMSEC) 1.09%, (RMSE External Calibration) 0.04%, and (RMSE External Cross-Validation) 0.48%. On the other hand, PCA successfully differentiated between pork and beef meatball formulations in market samples. Rahayu et al. (2018a, b) aimed to evaluate the effectiveness of FTIR spectroscopy combined with chemometrics (PCA and PLSR) for identifying and quantifying dog meat (DM) in beef meatballs (BM) as shown in Figs. 4.24 and 4.25. The study found that the combined frequency regions of 1782–1623 cm^{-1}, 1700–700 cm^{-1}, and 1485–659 cm^{-1}, with detrending treatment, provided optimal predictions for DM in BM. It has been reported that statistical evaluation showed the Folch extraction method produced higher R^2 values and lower RMSEC and RMSEP compared to the Bligh-Dyer method. PCA effectively classified meatballs containing DM from those made with other types of meat (Figs. 4.26 and 4.27). FTIR spectroscopy combined with multivariate analyses using PLSR and PCA proved to be a reliable and rapid method for screening DM in meatball products.

Rat meat in binary mixtures with beef in meatball formulations has also been investigated by Rahmania and Rohman (2015) using FTIR coupled with

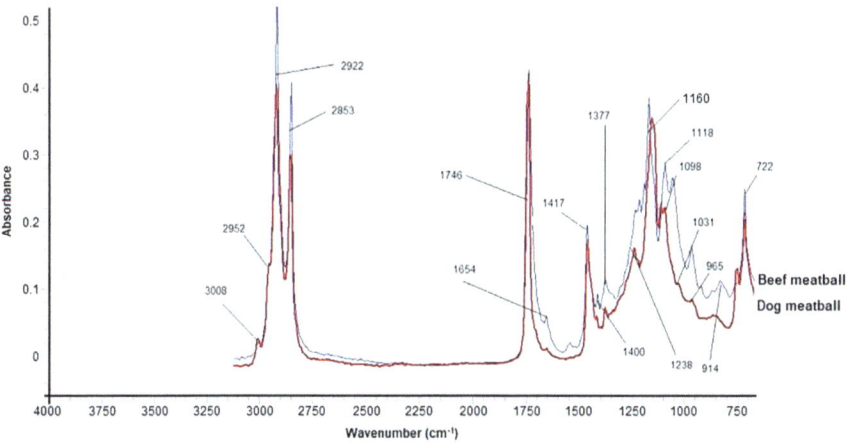

Fig. 4.24 FTIR spectra of lipid fraction extracted from 100% beef meatballs and 100% dog meatballs. (Rahayu et al. (2018a)

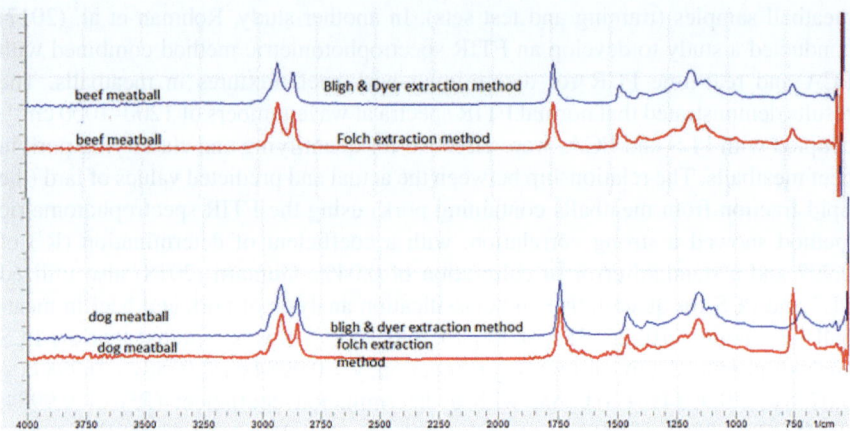

Fig. 4.25 FTIR spectra from dog meatballs (100% dog meat) and beef meatballs (100%) extracted with Bligh and Dyeras well as the Folch extraction method. (Rahayu et al. 2018b)

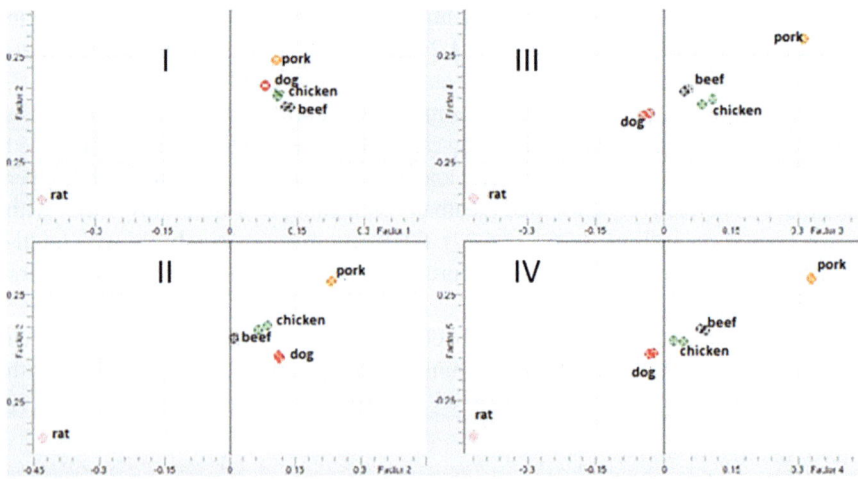

Fig. 4.26 PCA score plot of fats extracted from meat(pork, dog, chicken, beef, and rat), representing the projection of samples defined by the 1st and 2nd PC (I), 2nd and 3rd PC (II), 3rd and 4th PC (III), and 4th and 5th PC (IV). (Rahayu et al. 2018b)

chemometric models (PCA and PLS). After optimizing frequency regions in the mid-IR spectrum, the region of 750–1000 cm^{-1} was selected for PLS and PCA modeling. In another study, Guntarti and Prativi (2017) prepared meatballs with varying rat meat concentrations (0%, 25%, 35%, 65%, 75%, and 100%) for qualitative and quantitative identification of rat meat in meatballs. After extraction of lipids from meatballs, FTIR spectra were selected (wavenumber 750–1600 cm^{-1}) for PLS and PCA models. The PLS calibration model showed a strong correlation ($R^2 = 0.9941$) with a RMSEC of 1.63%, and validation yielded an RMSECV of 1.79% and an RMSEP of 2.60%. Meanwhile, PCA effectively distinguished between beef and rat

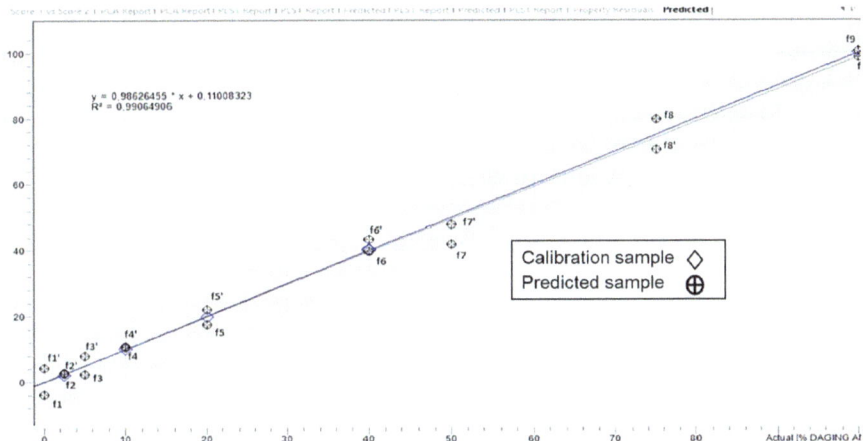

Fig. 4.27 Relationship between actual values of dog's fat from Folch extraction method (x-axis) and FTIR calculated values (y-axis) modeled using PLS wavenumbers of 1700–700 cm-1, f1 and f1' (0%), f2 and f2' (2.5%), f3 and f3' (5%), f4 and f4' (10%), f5 and f5' (20%), f6 and f6' (40%), f7 and f7' (50%), f8 and f8' (75%), f9 and f9' (100%). (Rahayu et al. 2018b)

fat in the meatballs, confirming its utility for grouping. The method was successfully applied to quantify rat meat in commercial meatball samples. In a recent study, Lestari et al. (2022) investigated three lipid extraction methods, Bligh and Dyer, Folch, and Soxhlet, for detecting rat meat adulteration in beef meatballs using FTIR spectroscopy. The FTIR spectra at wavenumber regions of 3100–800 cm^{-1} were utilized for chemometric modeling (LDA) and successfully classified lipid components from beef and rat meatballs with 100% accuracy across all extraction methods. Rat meat adulteration (Halal authentication) was also effectively quantified using PLS and PCR under optimized conditions. Wild boar meat in beef meatballs has also been analyzed by Guntarti et al. (2015) and Ahda et al. (2020) using FTIR spectroscopy combined with chemometric methods and PCA to quantify and classify wild boar and beef in the meatball formulations. FTIR spectroscopy, combined with reported PLS and PCA at the wavenumber range of 1250–1000 cm^{-1}, proved effective for the quantitative analysis of wild boar meat in meatballs Guntarti et al. (2015). While Ahda et al. (2020) reported that PCA effectively distinguished between wild boar and beef meatballs (Figs. 4.28 and 4.29) prepared with varying concentrations (0–100%) at specific spectral regions, 999–1481 cm^{-1} and 1650–1793 cm^{-1} (processed at the first derivative) as key markers for wild boar detection.

Detection of animal fat mixtures in meatballs using FTIR spectroscopy has been investigated by Fajriati et al. (2021), focusing on the unique fat markers in mixtures containing lard, chicken fat, and rat fat. The fat samples were prepared by varying the ratios of chicken fat (100%) mixed with 1%, 10%, and 20% lard or rat fat. Meatball formulations included 0%, 5%, and 90% rat meat. FTIR analysis revealed that increasing concentrations of lard and rat fat led to higher absorption at wavenumbers 3371 cm^{-1}, 3332 cm^{-1}, 2337 cm^{-1}, and 1743 cm^{-1}. These spectral changes were associated with the functional groups characteristic of animal fats. The study

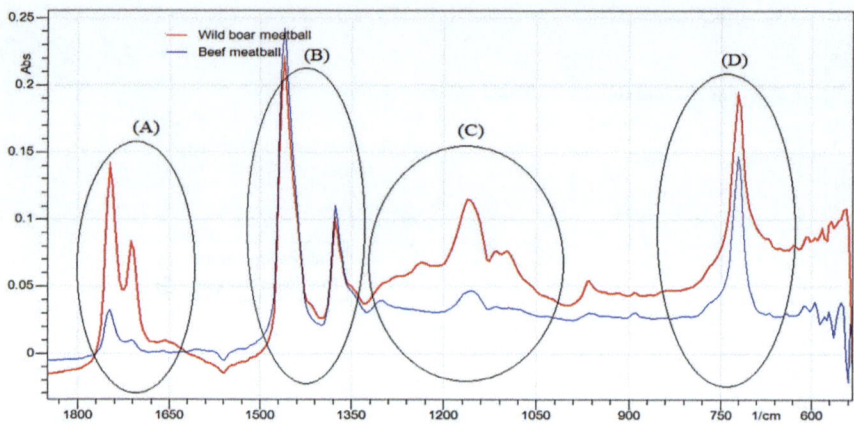

Fig. 4.28 FTIR spectrum from wild boar meatball and beef meatball at wavenumber 500–1850 cm^{-1}. (Ahda et al. 2020)

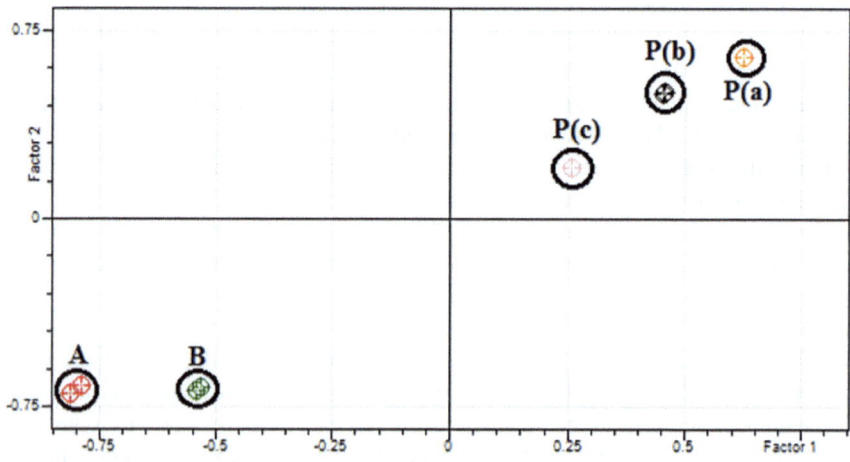

Fig. 4.29 Detection of meatball sample (P(a), P(b), and P(c)) from commercial product. (A) Beef meatball and (B) Wild boar meatball (Ahda et al. 2020)

demonstrated that the distinct properties of animal fats could be consistently identified using FTIR spectroscopy, providing a reliable method for analyzing fat mixtures in meatballs. Recently, Nurani et al. (2022) explored the use of FTIR spectroscopy combined with chemometrics to classify and quantify pork in roasted beef meatballs. For data acquisition, FTIR spectra of pure beef meatballs and beef meatballs adulterated with pork at various concentrations were recorded in the mid-IR region (4000–400 cm^{-1}). Under optimal conditions and wavenumber (1100–900 cm^{-1}) using PLS, the relationship between actual pork concentrations and FTIR-predicted values was highly accurate, with a coefficient of determination (R^2) of 0.99965 and a RMSEC of 0.7137%. PCA successfully differentiated between pure beef and pork meatballs (market samples).

4.7.3 Sausages

Sausage is another type of meat product that is typically produced from chicken meat or beef. Sausage is also susceptible to adulteration with other, less expensive meats for economic considerations. The detection of other meats by visual inspection is inherently challenging because they are combined with various ingredients (Habiba et al. 2021). According to the literature, FTIR spectroscopy combined with chemometrics has been successfully employed to detect and authenticate sausage-based products. This method has been effectively utilized to identify lard adulteration in sausage as well as the presence of rat and dog meat in beef sausages. Table 4.9 presents data from some of the published studies on (sausage-based products) Halal authentication using FTIR spectroscopy.

Discrimination of pork in Chinese ham sausages was developed by Xu et al. (2012) using FTIR spectroscopy combined with chemometrics. Transmittance spectra of Chinese ham sausages (78 non-Halal and 73 Halal) were recorded, ranging from 400 to 4000 cm^{-1}. The impact of various data preprocessing techniques, such as smoothing, derivative calculations (Fig. 4.30), and standard normal variate (SNV) on least squares support vector machine (LS-SVM) and PLSDA, was investigated. The optimal models exhibited sensitivity and specificity values of 0.913 and

Table 4.9 The application of FTIR spectroscopy combined with chemometrics for authentication analysis of sausage products

Adulterated meat/food products	Meat adulterants	Spectral range	Spectral treatment and chemometrics	References
Beef and chicken sausages	Pork	4000–400 cm^{-1}	PLSDA	Xu et al. (2012)
Beef/veal sausages	Pork	6028–5480 cm^{-1}	PCA and SVM	Schmutzler et al. (2015)
Beef sausages	Rat meat	1800–750 cm^{-1}	PLSR	Pebriana et al. (2017)
	Wild lard and cow fat	1250–900 cm^{-1}	PCA	Sari and Guntarti (2018)
	Dog fat	1124–688 cm^{-1}	PLS and PCA	Guntarti and Purbowati 2019
	Lard	1200–1000 cm^{-1}	PCA and PLS	Guntarti et al. (2019a)
	Lard/pork	1000–791 cm^{-1}, And 1070–796 cm^{-1}	PLS	Guntarti et al. (2019b)
	Dog meat	1800–850 cm^{-1}	PCA and PLS	Wirnawati et al. (2023)
	Pork	1200–1000 cm^{-1}	PCA	Ahda et al. (2023)
	Rat meat	3100–700 cm^{-1}	PCR and PLS	Lestari et al. (2024)

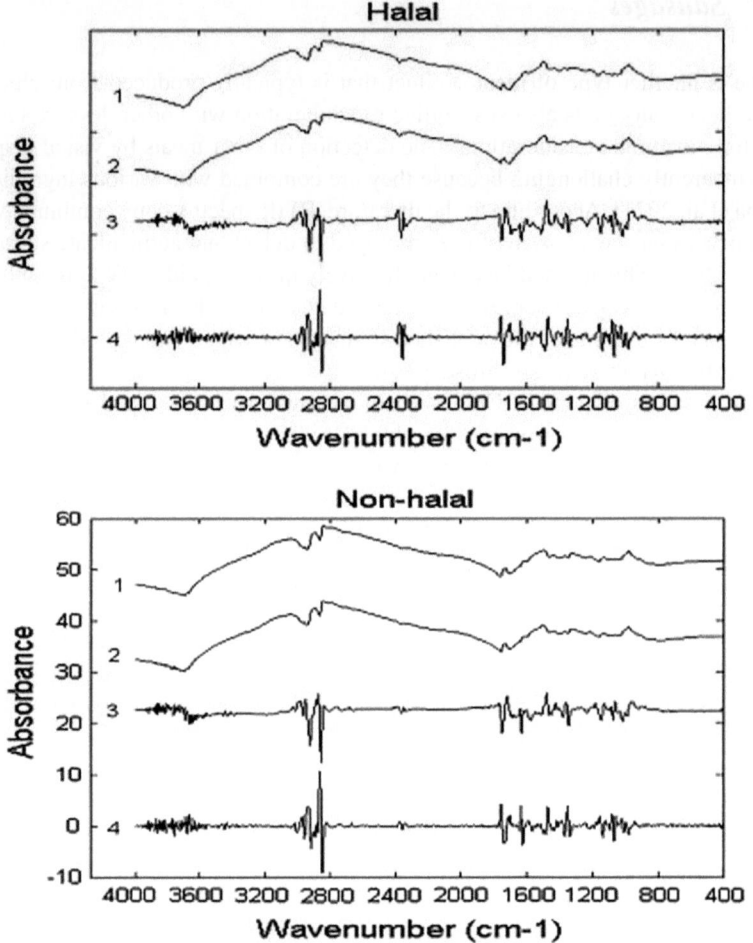

Fig. 4.30 Average spectra of Halal and non-Halal ham sausages preprocessed by smoothing (1), SNV (2), first-order derivative (3), and second derivative (4). (Xu et al. 2012)

0.929 for PLSDA using SNV spectra and 0.957 and 0.929 for LS-SVM employing 2nd derivative spectra, in each case (Figs. 4.31 and 4.32).

In 2015, Schmutzler and coworkers developed the NIR-based approach for identifying the pork adulteration in the meat and fat part of veal sausages. The sausages were independently prepared using a commercial veal product. Within the wave number range of 6028–5480 cm^{-1}, adulterations containing pork and pork fat up to 50% (in 10% steps) were examined. PCA models were developed for each setup, incorporating prior data pre-treatment procedures such as wavelength selection, scattering corrections, and spectral data derivatives. PCA scores served as input data for the validation and classification processes of SVM. It was reported that the lowest level of contamination (10%) could be detected in meat and fat adulterated

4.7 FTIR Spectroscopy for Halal Authentication

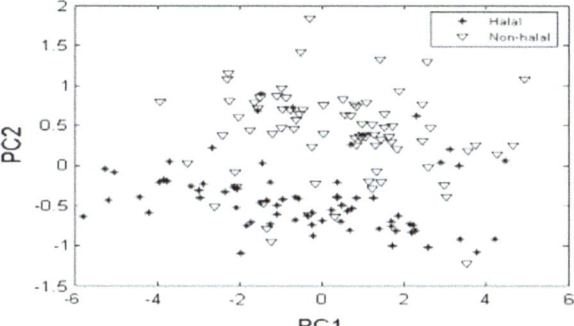

Fig. 4.31 PCA plot of raw FTIR spectra of Halal and non-Halal ham sausages. (Xu et al. 2012)

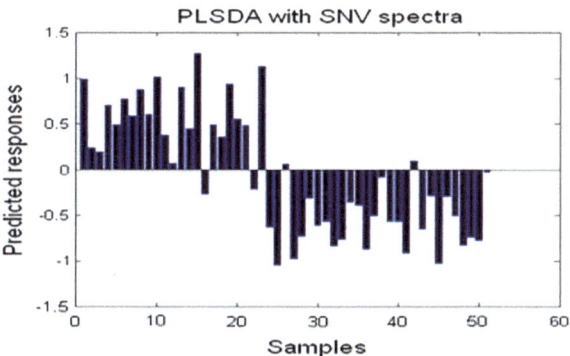

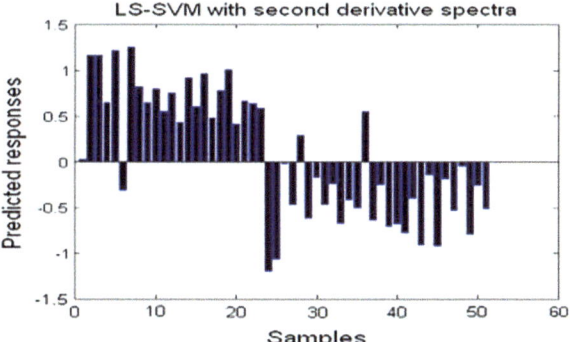

Fig. 4.32 The predicted response values are based on the best linear PLSDA model and nonlinear LS-SVM models. Samples 1–23 are Halal (positive) samples, and 24–51 are non-Halal (negative) samples. (Xu et al. 2012)

samples using industrial fiber optics setup as well as laboratory setup. Moreover, Pebriana et al. (2017) reported the use of FTIR spectroscopy combined with chemometrics to analyze rat meat in sausages with three distinct lipid extraction procedures. The lipid was extracted from the sausage samples with the Soxhlet, Folch, and Bligh and Dyer methodologies. The extracted lipid was subsequently analyzed through FTIR spectroscopy in conjunction with chemometric techniques, specifically PCA and PLS calibration. Absorbance values within the wavenumbers range of 750–1800 cm^{-1} were utilized in PLS modeling for quantification and PCA for classification. PCA effectively classified rat meat and beef lipids obtained through the three lipid extraction methods. Sari and Guntarti (2018) employed ATR-FTIR in conjunction with PLS and PCA chemometrics to analyze wild lard and cow fat in sausage samples (Fig. 4.33). The PLS model was used for quantitative analysis in the wave number range of 1250–900 cm^{-1} (Fig. 4.34). The R^2 value generated

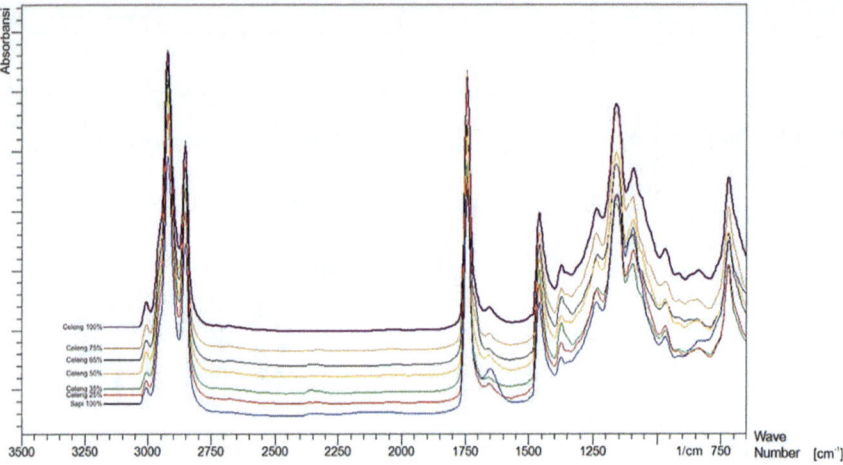

Fig. 4.33 FTIR spectrum of beef sausage lipid and wild boar sausage lipid in gradual concentration. (Sari and Guntarti 2018)

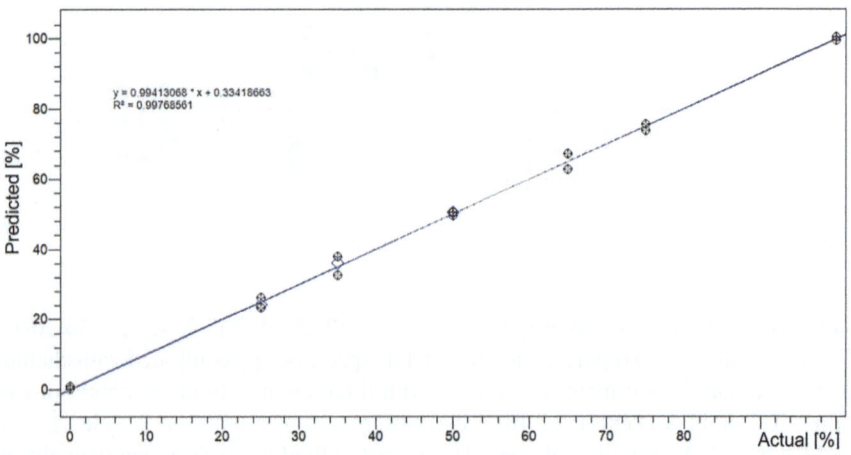

Fig. 4.34 The results of the PLS reference sample in gradual concentration. (Sari and Guntarti 2018)

4.7 FTIR Spectroscopy for Halal Authentication

by the multivariate calibration model was 0.998, RMSEC of 1.22%, RMSEP of 0.11%, and RMSECV of 2.68%. The PCA model was employed to classify the types of fat present in beef and wild boar sausage obtained from the market (Fig. 4.35). It was reported that the analysis of all samples revealed no traces of boar meat, as indicated by a profile consistent with that of beef sausage.

Guntarti et al. (Guntarti and Purbowati 2019; Guntarti et al. 2019a, b) reported three different studies on the determination and classification of dog fat in beef sausages, lard in beef sausages, and lard in grilled and steamed beef sausages using FTIR spectroscopy combined with chemometrics (PLS and PCA). For dog fat determination, different concentrations of dog fat were prepared (100%, 75%, 65%, 50%, 35%, and 25%), and five market samples were analyzed with optimum wave number range between 1124–688 cm^{-1}. The selected region provided strong differentiation, yielding a y = 0.9999 x + 0.0004 and R^2 equal to 0.9999. The results equation yielded RMSEC of 0.30%, RMSEP of 0.05%, and RMSECV of 0.05%. For the analysis of lard in beef sausages (Fig. 4.36), PLS and PCA were done on 1200–1000 cm^{-1}, yielding an R^2 value of 0.985 and RMSEC (2.094%). The RMSEP for external validation was 4.77%, while the RMSECV for internal validation was 5.12%. On the other hand, two different wave number ranges (1070–796 cm^{-1} and 1000–791 cm^{-1}) were selected for the determination of lard in grilled and steamed beef sausages using Horizon MBTM (Figs. 4.37 and 4.38). Samples of different concentrations (100%, 75%, 65%, 50%, 35%, 25%, and 0%) different concentration of lard and sausages were prepared. PLS analysis of steaming and grilled sausages yielded the following equation: (y = 0.9977x + 0.1166; R2 = 0.9977; RMSEC = 1.22%; RMSEP = 0.22%; and RMSECV = 1.26%) and (y = 0.9972x + 0.1379; R2 = 0.9972, RMSECV = 0.18%, RMSEP = 0.42%, and

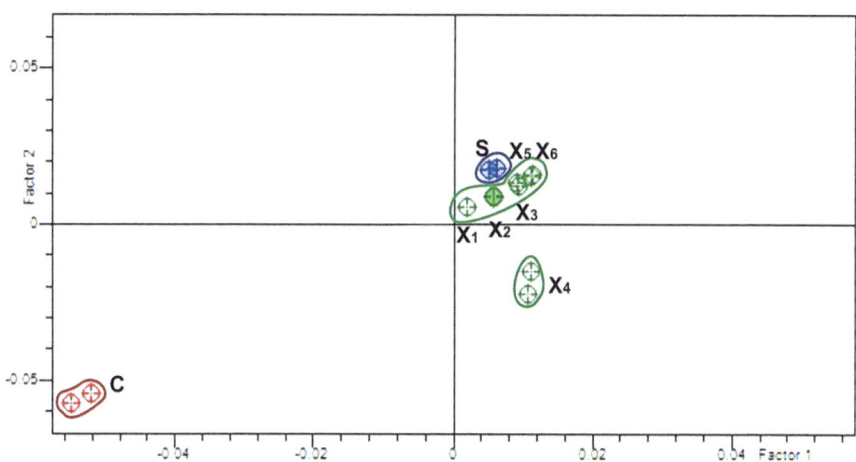

Fig. 4.35 The results of PCA analysis in a score plot. Note C (Wild Boar meat lipid), S (beef lipid), and X (commercial samples). (Sari and Guntarti 2018)

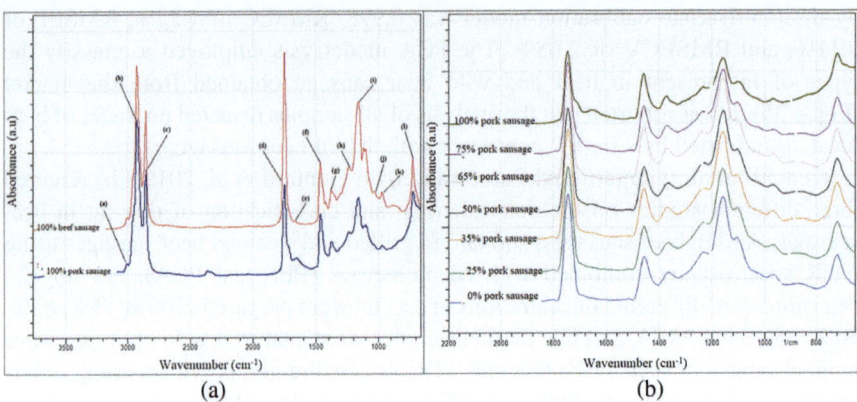

Fig. 4.36 (**a**) FTIR spectrum from pure beef sausage (100%) and pure pork sausage (100%). (**b**) Spectrum of FTIR sausage various concentrations reference (0–100%). Guntarti et al. (2019a)

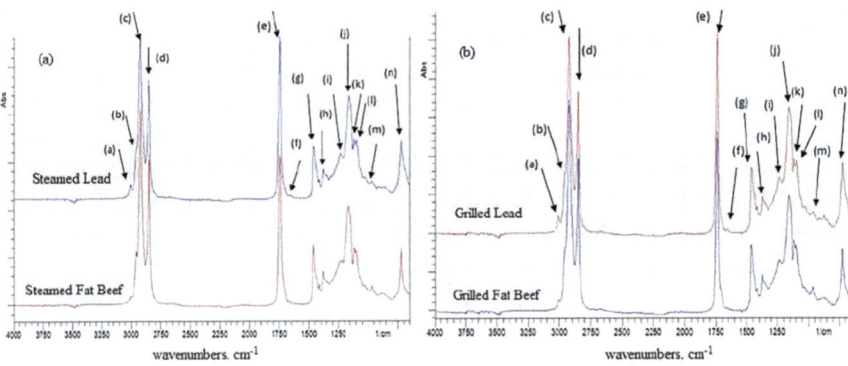

Fig. 4.37 (**a**) Spectra of steamed pork sausage fat of 100% and steamed beef sausage of 100%. (**b**) Spectra of grilled beef sausage fat of 100% and grilled pork sausage fat of 100%. (Guntarti et al. (2019b)

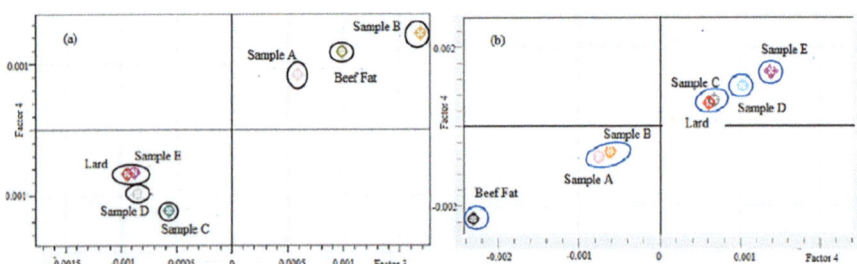

Fig. 4.38 The PCA result of products in the market of steamed sausage (**a**) and products in the market of grilled sausage (**b**). (Guntarti et al. (2019b)

RMSEC = 1.27%), respectively. PCA analysis's findings revealed the categorization of various quadrants between 100% beef and 100% pork sausage, respectively.

In another study, Wirnawati et al. (2023) identified and quantified fat derived from dog meat in beef sausage formulations. The lipid components in the whole mid-IR region 4000–650 cm^{-1} were investigated using TQ analysis and FTIR spectroscopy. DA was used for qualitative analysis, whereas PLS and PCR models were used for quantitative chemometrics analysis. The findings demonstrated that wave numbers 1800–850 cm^{-1} differentiated the beef sausages adulterated with dog meat using LDA. Lipid components obtained by Soxhlet extraction (beef sausages, dog sausages, and combinations of dog and beef sausages) were all successfully classified with 100% accuracy using LDA. Similarly, Ahda et al. (2023) detected pork in beef sausage using FTIR and GC-MS instruments. FTIR was used to perform DA of commercial sausage products at wavenumbers 1200–1000 cm^{-1}, whereas GC-MS was used to ensure the Halal-ness of sausages based on the FAME compositions. The data obtained were analyzed using PCA to differentiate beef sausage. Myristic acid, lauric acid, and palmitoleic acid were fatty acids that set pork sausage apart from beef sausage, according to the loading plot. Based on these findings, FTIR and GC-MS, in conjunction with chemometrics, successfully assessed Halal authentication of sausage products (pork sausage and beef sausage). A recent study by Lestari et al. (2024) analyzed rat meat adulteration in beef sausages by using FTIR-ATR spectroscopy in combination with chemometrics. Lipid components of sausages were extracted using three different methods (Soxhlet, Folch, and Bligh and Dyer). Samples were made by combining beef with rat meat adulterant at concentrations ranging from 0–100%, respectively. The extracted lipid components were discriminated with a 100% accuracy level using absorbance value in wavenumber regions of 3100–700 cm^{-1}. Using multivariate calibration (PCR and PLS) under ideal conditions, rat meat adulterated in beef sausages was effectively ascertained.

4.7.4 "Rambak" Cracker

Rambak represents a traditional food product from Indonesia, typically made from the skins of cows and buffaloes. Rambak exhibits a broad range of applications when combined with many different food items, and it is particularly prevalent in the markets of Indonesia and Malaysia (Mursyidi 2013). Pig skin, as opposed to cow or buffalo skin, is one of the several sources that can be utilized in the production of Rambak. Distinguishing between Rambak made from pig skin and those made from other animal skins presents significant challenges, as the final product exhibits striking similarities. Various studies have focused on the utilization of FTIR as a rapid method for the authentication of beef jerky (Table 4.10).

Erwanto et al. (2016) and Muttaqien et al. (2016) both investigated the classification and quantification of lard and different animal skin in "rambak" crackers using

Table 4.10 The application of FTIR spectroscopy combined with chemometrics for authentication analysis of skin and animal fats

Meat/animal fat products	Adulterants fat/products	Spectral range	Spectral treatment and chemometrics	References
"Rambak" crackers	Lard	1200–1000 cm^{-1}	PLS	Erwanto et al. (2016)
"Rambak" crackers containing buffalo skin			PCA and PLS	Muttaqien et al. (2016)
Cowhide and artificial PVC-based leather	Pig skin	1534, 1736, 1277, 817, 1723, and 744 cm^{-1}	Simple regression analysis	Syabani et al. (2023)
Beef jerky	Pork	1500–600 cm^{-1}	LDA, SIMCA, and SVM	Kuswandi et al. (2015a, b)
Animal body fats (lamb, cow, and chicken)	Lard	3008–3000, 1418–1417, 1385–1370, 1126–1085, and 968–965 cm^{-1}	Simple regression analysis	Che Man and Mirghani (2001)
Animal body fats (lamb, cow, and chicken)		3010–2000, 1220–1095, and 968–965 cm^{-1}	PLS	Jaswir et al. (2003)
Animal fats	Bovine, poultry, lamb, pig, and fish oil	3600–200 cm^{-1}	PCA and PLS-DA	Abbas et al. (2009)
Animal body fats (lamb, cow, and chicken)	Lard	3300–650 and 1500–900 cm^{-1}	PLS-DA	Rohman and Che Man (2010)
Animal body fats (lamb, cow, and chicken fat)		4000–650 cm^{-1}	PLS	Rohman et al. (2011a, b)
Animal fats (beef, chicken, and mutton fat)			PCA	Naquiah et al. (2017)
Chicken fat		1236 and 3007 cm^{-1}	PCA	Saputra et al. (2018)
Edible fats, chicken fat, mutton fat, tallow, palm-based		2800–3200 cm^{-1}	PCA and LDA	Salleh et al. (2018)

4.7 FTIR Spectroscopy for Halal Authentication

FTIR spectrophotometer and the chemometrics of PCA and PLS (Fig. 4.39). Lard in "rambak" crackers was effectively quantified and classified using FTIR spectroscopy in wavenumber ranges of 1200–1000 cm^{-1}. The findings demonstrated a correlation of 0.96 between the expected and actual values of pig skin in Rambak (RMSEP of 1.10 and RMSEC of 2.56). Pig, buffalo, and commercial Rambak skin types are all successfully classified by the PCA models (Fig. 4.40). Recently, Syabani et al. (2023) investigated the authentication of Halal leather by employing various analytical techniques such as FTIR Spectroscopy, SEM, and DSC. FTIR study revealed major absorption bands: pig skin had a clear peak at 1534 cm^{-1}, cowhide had prominent peaks at 1736, 1277, and 817 cm^{-1}, and artificial PVC-based leather showed distinctive stretching at 1723 and 744 cm^{-1}. Various types of leather can be distinguished through the rapid and precise analytical

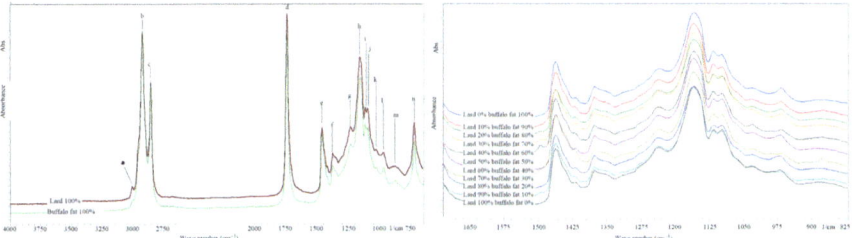

Fig. 4.39 (**a**) FTIR spectra of lipid fraction extracted from Rambak cracker containing 100% buffalo skin (buffalo fat) and 100% pig skin (lard). (**b**) Overlay spectra of lard mixed into buffalo fat at a concentration range of 0–100.0% (v/v) at mid-IR region (4000–650 cm^{-1}). (Muttaqien et al. 2016)

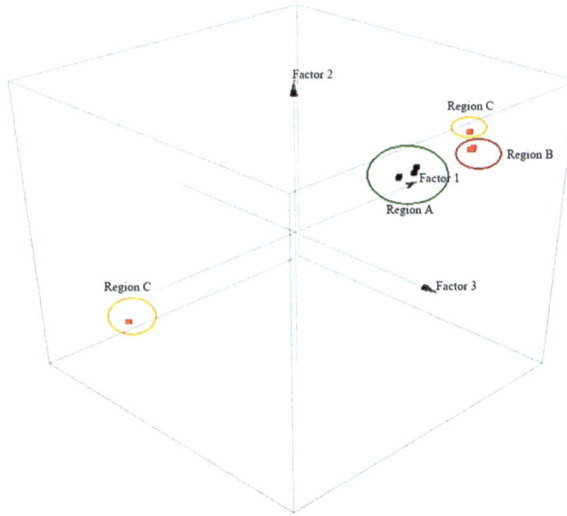

Fig. 4.40 PCA score plot (3 Dimension), expressed as the first principal component (PC1) and second principal component (PC2) for classification of Rambak with lard, buffalo, and commercial sample. (Muttaqien et al. 2016)

technique offered by FTIR. Although the spectra of pig skin and cowhide leathers exhibited similarities, practical differentiation can be achieved through the observed variations in vibration intensity.

4.7.5 Beef Jerky (Dendeng)

Beef jerky, commonly referred to as dendeng, is primarily derived from beef and is a favored delicacy in Malaysia and Indonesia. The components of beef jerky include brown sugar, salt, specific spices, and beef. The beef utilized in the production of jerky is prone to adulteration with pork due to the lower cost of pork compared to beef (Regenstein et al. 2003). Furthermore, detecting any trace of pork in cooked beef jerky through visual inspection poses significant challenges, as the appearance closely resembles that of the beef, making differentiation quite difficult (Table 4.10). Kuswandi et al. (2015a, b) used FTIR-ATR spectroscopy and chemometric modeling to detect pork adulteration in beef jerky. Samples with 5–80% pork were analyzed across four spectral data sets, with LDA, SIMCA, and SVM applied for classification. The LDA model using the full spectral region achieved 100% accuracy in identifying adulteration and recording spectral data within the 4000–700 cm^{-1} ranges, respectively.

4.7.6 Body Fats of Various Animals

Meat and body fats from various animals exhibit distinct biochemical compositions, making their authentication crucial for ensuring food quality, safety, and compliance with dietary laws such as Halal and Kosher. FTIR spectroscopy has emerged as a powerful tool for distinguishing these animal-derived products based on their unique molecular fingerprints. The composition of meat primarily consists of proteins, lipids, water, and trace amounts of carbohydrates and minerals. Body fats, on the other hand, are composed mainly of triglycerides, with variations in fatty acid profiles depending on the animal species. For instance, lard (from pigs) contains higher levels of monounsaturated fatty acids, while beef and lamb fats are richer in saturated fats. Chicken fat typically has a higher proportion of polyunsaturated fatty acids compared to mammalian fats. Chemometric methods like PCA and PLSR further enhance FTIR's discriminative power by identifying patterns and correlations within complex spectral data. Studies have shown that FTIR combined with chemometrics can accurately detect lard adulteration in animal fats such as (bovine, mutton fat, poultry, lamb, pig, fish oils, and cow body fat), ensuring the authenticity of Halal and Kosher products (Table 4.10).

Che Man and Mirghani (2001) used FTIR spectroscopy for the quantification and detection of lard adulteration in lamb, cow body fat, and chicken. Spectral differences between lard blends and pure fats were identified with key absorption bands 3009–3000 cm^{-1}, 1418–1370 cm^{-1}, and 1126–966 cm^{-1}, regions serving as qualitative markers. A semi-quantitative approach was developed using regression equations based on frequency shifts and absorbance intensities, achieving high coefficients of determination (R^2 = 0.9233–0.9831) with low standard errors. The study demonstrated that absorbance at 966.22 cm^{-1} was highly correlated with lard content in lard-cow fat blends, while 3005.6 cm^{-1} was useful for lard-chicken fat blends. A few years later, another study by Jaswir et al. (2003) investigated the quantification and detection of lard adulteration in cow body fat and mutton using FTIR spectroscopy combined with chemometrics. The spectral analysis identified absorption regions at 3010–3000, 1220–1095, and 968–965 cm^{-1} for mutton lard blends while 1419–1414, 968–965 cm^{-1} for cow blends. PLS regression models achieved high accuracy (R^2 = 0.9866 and 0.9749, SE = 2.01 and 1.86, respectively), confirming FTIR's capability for lard quantification. Similarly, Abbas et al. (2009) assessed the discrimination of animal fats (bovine, poultry, lamb, pig, and fish oils) using FTIR combined with PLS-DA and PCA within the range 3600–200 cm^{-1}, respectively. PCA effectively grouped fats, showing significant separation of fish oil due to its high polyunsaturated content. PLS-DA models achieved sensitivity and specificity values of 0.958 and 0.914. The investigation of lard (LD) in mixes containing body fats of chicken (Ch-BF), lamb (LBF), and cow (Cow-BF) has been reported by Rohman and Che Man (2010) and Rohman et al. (2011a, b). The developed FTIR spectroscopic method, in conjunction with chemometrics (PLS and DA), was employed for the body fat analysis. It was suggested that FTIR spectra qualitatively differ from one another in order to distinguish between lard and its mixtures. The Mahalanobis distance concept in the whole mid-IR spectrum (3300–650 cm^{-1}) was effectively used by DA to distinguish between lard and mixes containing LBF, Cow-BF, and Ch-BF. For the determination of the percentage of lard in LBF, Cow-BF, and Ch-BF, a specific fingerprint region (1500–900 cm^{-1}) was selected, which showed R^2 value >0.99. Naquiah et al. (2017) differentiate lard, namely, lard stearin (LS) and lard olein (LO), from other frequent animal fats (beef, chicken, and mutton fat) using FTIR, DSC, GC-MS, and elemental analyzer–isotope ratio mass spectrometry (EA-IRMS). The FTIR spectral analysis of LO and LS compared to lard, chicken fat (ChF), beef fat (BF), and mutton fat (MF) revealed similarities in overall spectral appearance, with key differences in band intensity and frequency shifts (3008, 2918, 2853, 1748.5, 1457, 1378, 1220, and 1163 cm^{-1}). The PCA model, with the help of GCMS data (fatty acids), successfully discriminated between the LO and LS. A study conducted by Saputra et al. (2018) used a wavelength biomarker derived from FTIR in conjunction with PCA analysis to detect pig adulteration in various fat sample mixtures and specific foods. Comparing Halal food (HF), Non-Halal Food A (NHFA), and Non-Halal Food B (NHFB) fat samples to other fat samples, they were clearly distinguished at a noticeable distance at wavelengths 1236 and 3007 cm^{-1} along the spectrum. HF was found closer to ChF, suggesting that the sample may have included chicken fat, but the previous two samples, NHFA and NHFB, were near PF (Pig Fat), indicating that they had pork fat. The study by Salleh et al. (2018) investigated the differentiation of edible fats, ChF, MF, tallow, palm-based, and lard at various heating treatments using FTIR and multivariate analysis at wave numbers 2800–3200 cm^{-1}. PCA, LDA, and k-mean cluster analysis (k-mean CA) were employed to evaluate the effectiveness

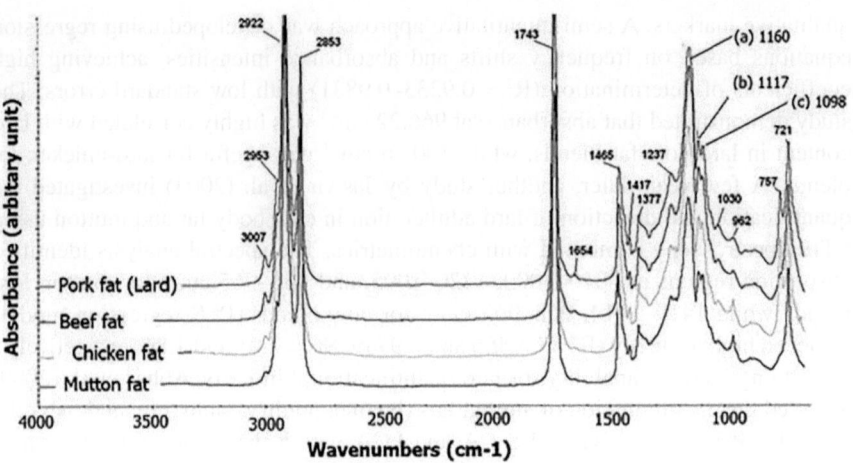

Fig. 4.41 FTIR spectra of lard, beef fat, chicken fat, and mutton fat at wavenumbers of 4000–650 cm^{-1}. (Rohman and Windarsih 2020)

of each method in distinguishing the fats following the heating treatments. The ability to discriminate heated fats based on their source was discovered when PCA and k-mean CA were combined. In its group, the LDA technique effectively classified 80.5% of the fat samples. A representative FTIR spectrum of different meat fats is shown in Fig. 4.41.

4.7.7 Oil and Fats

The legitimate origin of edible oils is essential for consumer health, economic considerations, and religious factors. A rapid, reliable, and accurate technique is necessary to identify edible oils, given the varying costs of different oils. FT-IR spectroscopy, in conjunction with chemometrics, is gaining prominence for the Halal verification of edible oils. A variety of FT-IR spectroscopic methods (ATR and transmission) have been established to authenticate various edible oils, encompassing seed oils, nut oils, and fruit oils. Table 4.11 presents a summary of FT-IR research on the authenticity of oils (Halal).

Guillén and Cabo (1997) characterized edible oils (OO, VOO, EVOO, sesame oil, virgin sesame oil, refined sunflower oil, refined corn seed oil, sunflower, and corn oils) and lard using transmission FTIR spectroscopy, focusing on the fingerprint region (1400–1000 cm^{-1}) to establish relationships between spectral band frequencies and fat composition. Spectra were recorded from 4000–500 cm^{-1} using KBr-disc methodology. Significant correlations were found between the unsaturated fatty acids and band frequencies at 3006 cm^{-1}, 1400 cm^{-1}, and 1097 cm^{-1}, enabling quantitative prediction of oil composition. A study by Yang et al. (2005) investigated the discrimination of edible oils and fats using FTIR, FT-Raman,

Table 4.11 The application of FTIR spectroscopy combined with chemometrics for authentication analysis of oils

Adulterated fats and oils	Adulterants	Spectral range	Chemometrics techniques	References
Olive oil, extra-virgin olive oil, virgin olive oil, sesame oil, virgin sesame oil, refined sunflower oil, refined corn seed oil, sunflower, and corn oils	Lard	1400–1000 cm^{-1}	–	Guillén and Cabo (1997)
Canola, coconut, soybean, olive, safflower, corn, pea, cod liver, and butter		400 and 4000 cm^{-1} 1400 and 1800 cm^{-1}	LDA and CVA	Yang et al. (2005)
Walnut, canola, corn, cod liver, sunflower, grape seed, rice bran, palm, pumpkin seed, extra virgin olive, soybean, sesame, and virgin coconut oils		2922 and 1464 cm^{-1}	PCA	Che Man et al. (2011a, b)
Extra virgin olive oil, corn oil, canola oil, and sunflower oil		1500–1000 cm^{-1}	PLSR	Che Man and Rohman (2011)
Virgin coconut oil		3020–3000 and 1120–1000 cm^{-1}	PLS and DA	Mansor et al. (2011)
Vegetable oils (canola oil, corn oil, extra virgin olive oil, soybean oil, and sunflower oil)		1500–1000 cm^{-1}	PLS, PCR, and DA	Rohman et al. (2011a, b)
Palm oil		1480–085 cm^{-1}	–	Rohman et al. (2012)
		3100–1050 cm^{-1}	PLS	Che Man et al. (2014)
Crude palm oil		1481–999 and 1793–1650 cm^{-1}	PLS	Ahda et al. (2016)
Soybean oil		1226 and 1111 cm^{-1}	PLS, PCA	Utami et al. (2018)
Palmolein oil		3006 and 1117 cm^{-1}	PLSR	Sim et al. (2018)
Sunflower, canola, coconut, olive, and mustard oil		1078 and 1246 cm^{-1}	PLS	Munir et al. (2019)
Cooking oil		1745, 1163, and 722 cm^{-1},	PCA and PLS-DA	Ahda et al. (2024)
Horse oil, butter, and sheep oil		4000–400 cm^{-1}	CNN and RNN	Kuang et al. (2025)

and FT-NIR spectroscopy combined with chemometric techniques. In this study, ten different oils and fats (canola, coconut, olive, soybean, safflower, corn pea, cod liver, lard, and butter) were analyzed between wavenumbers at 400 and 4000 cm^{-1}, 1400 and 1800 cm^{-1}. Canonical Variate Analysis (CVA) and LDA were applied for classification. Among the investigated techniques, FTIR achieved the highest accuracy (98%), followed by FT-Raman (94%) and FT-NIR (93%), respectively. Similarly, FTIR spectroscopy spectra at MIR region (4000–650 cm^{-1}) with chemometric method use for the differentiated 16 edible fats and oils (walnut, canola, corn, cod liver, sunflower, grape seed, rice bran, palm, pumpkin seed, EVO, soybean, sesame, and VCO) with lard by Che Man et al. (2011a, b). An eigenvalue of around 90% was attained by utilizing four PCs in PCA. PC1 represented 44.1% of the variance, whilst PC2 accounted for 30.2% of the variance. The primary frequency regions affecting the differentiation of lard from other assessed oils and fats according to PC1 were 2852, followed by 2922 and 1464 cm^{-1}, respectively. Additionally, CA categorized lard into its group according to Euclidean distance. In another study, Che Man and Rohman (2011) evaluated the adulteration of vegetable oils (EVOO, corn oil, canola oil, and sunflower oil) with lard by using spectroscopy, chromatography, thermogravimetry, and Electronic Nose (EN) techniques. For quantitative analysis, it was suggested that FTIR in the 1500–1000 cm^{-1} fingerprint region was effective for distinguishing lard from vegetable oils, achieving a high correlation ($R^2 > 0.99$) using PLSR. The study also highlights FTIR spectroscopy as a robust, non-destructive tool for food authentication, with chromatographic and thermal methods serving as complementary techniques for Halal verification and regulatory enforcement. Mansor et al. (2011) from the same research group analyzed the presence of lard in VCO with both techniques FTIR and GC techniques coupled with PLS and DA for the quantification and classification of LD in VCO. The findings indicated that PLS effectively predicted the LD content in VCO using the equation of y = 0.999 × +0.006, demonstrating a correlation between the actual LD value (x) and the FTIR predicted value (y) with a R^2 of 0.9990 in the frequency ranges of 3020–3000 cm^{-1} and 1120–1000 cm^{-1}. DA can be categorized as VCO and adulterated with LD utilizing the FTIR spectra at the identical frequency areas employed in quantification. PLS and DA were employed by Rohman et al. (2011a, b) to measure and categorize the amount of lard in some vegetable oils (corn oil, canola oil, sunflower oil, soybean oil, and EVOO) at a fingerprint region (1500–1000 cm^{-1}). It was highlighted that lard, in combination with vegetable oils, can be quantified individually using PLS calibration with its first derivatives or with normal spectra.

Previously, many authors have studied the detection of lard adulteration in palm oil using FTIR spectroscopy in conjunction with chemometric modeling (Rohman et al. 2012; Ahda and Safitri 2016; Sim et al. 2018). Rohman et al. (2012) found a strong linear correlation ($R^2 = 0.998$) between the actual and predicted lard values, with SECV = 2.87% (v/v) being marginally higher than RMSEC = 1.69% (v/v) when measuring the lard content in palm oil at frequencies of 1480 and 1085 cm^{-1}. Similarly, Ahda et al. (2016) further refined lard detection in crude palm oil in combination with the specific fingerprint regions 1481–999 cm^{-1} and 1749–1650 cm^{-1}.

The actual value and predicted value at the combination of wave numbers produce R^2 value of ~0.998 with the lowest detection error due to smaller RMSEC and RMSECV values of 1.291% (v/v), 0.838% (v/v), respectively. On the other hand, Sim et al. (2018) expanded this approach by predicting lard adulteration in palmolein oil. Spectral data were recorded in the 4000–525 cm^{-1} range (Fig. 4.42), with key absorption bands identified as marker bands 3006, 2852, 1117, 1236, and 1159 cm^{-1} (Fig. 4.43). PLSR, among the regression models (Fig. 4.44), demonstrated superior predictive performance, achieving a %RMSE values of 16.03 and 13.26% for the bands at 3006 cm^{-1} and 1117 cm^{-1}, respectively.

Che Man and coworkers (2014) detected the presence of palm oil adulteration with lard in pre-fried French fries. An FTIR calibration model with key features of interest was developed (spectral region of 3100–1050 cm^{-1}) to predict the concentration of lard in a blend of lard and palm oil. The model yielded a coefficient of determination (R^2) 0.9791, with a detection limit of 0.5%. On the other hand, the calibration error (RMSEC) was 0.979%, while the SECV was 2.45%. Lard in seasoning oil of imported instant noodle products using ATR-FTIR combined with chemometrics (PCA and PLS) has been reported by Utami et al. (2018). The authors revealed two typical wave numbers of lard at 1226 and 1111 cm^{-1} in the IR spectrum. With wave numbers between 2924 and 717 cm^{-1}, the PLS model yielded $R^2 = 0.992103$, RMSEP = 0.0072, and RMSEC = 1998, respectively. PCA generated a score plot demonstrating that the pork and soybean oil diverged into distinct

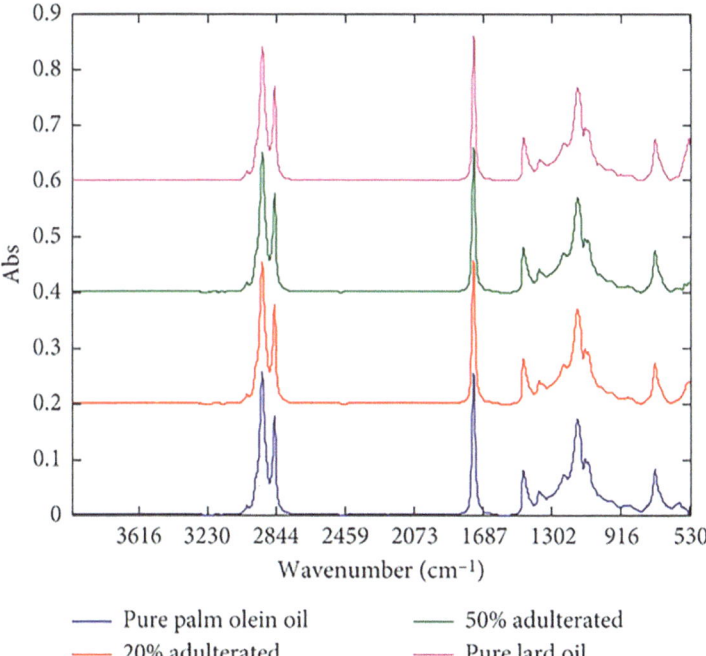

Fig. 4.42 The infrared spectra profile of pure and adulterated oil. (Sim et al. 2018)

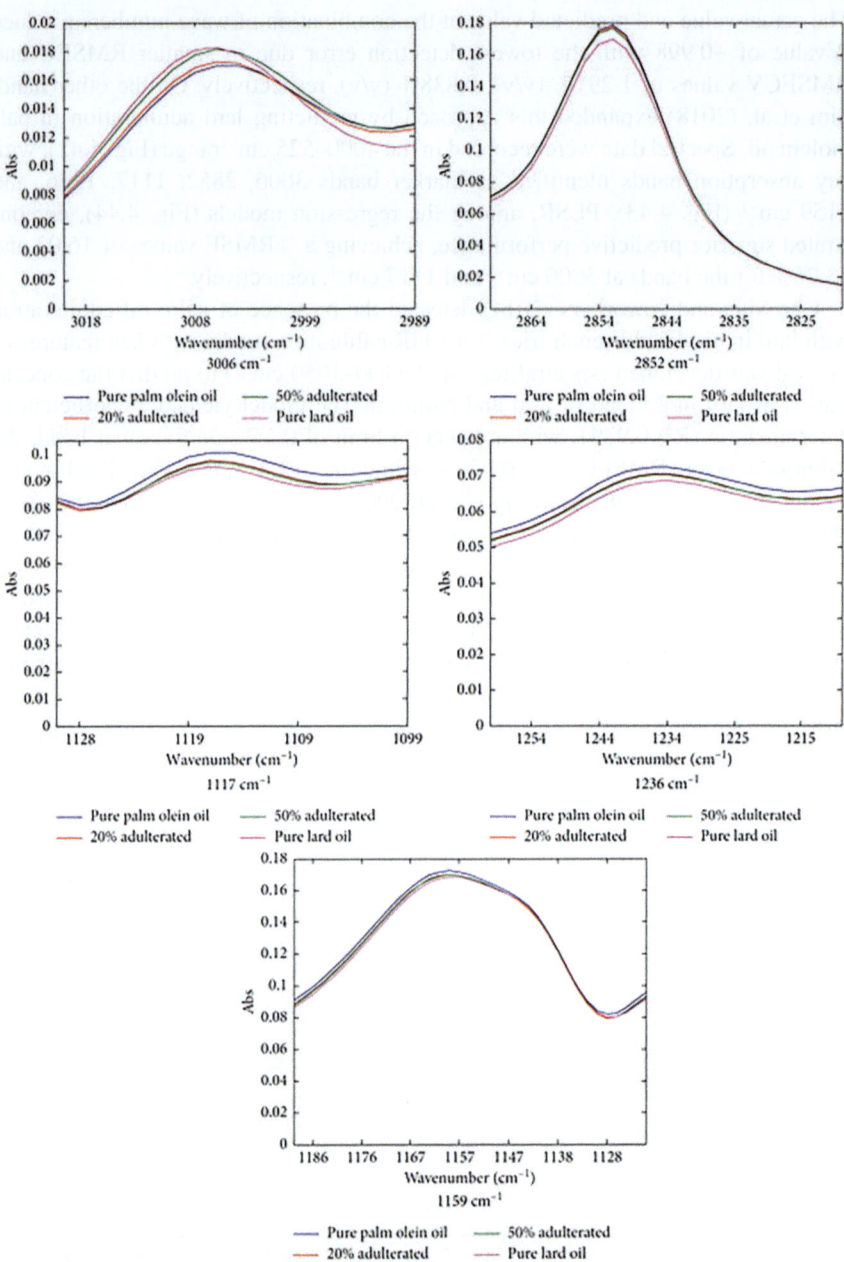

Fig. 4.43 The spectral region of five variables with the most significant discriminatory ability. (Sim et al. 2018)

4.7 FTIR Spectroscopy for Halal Authentication

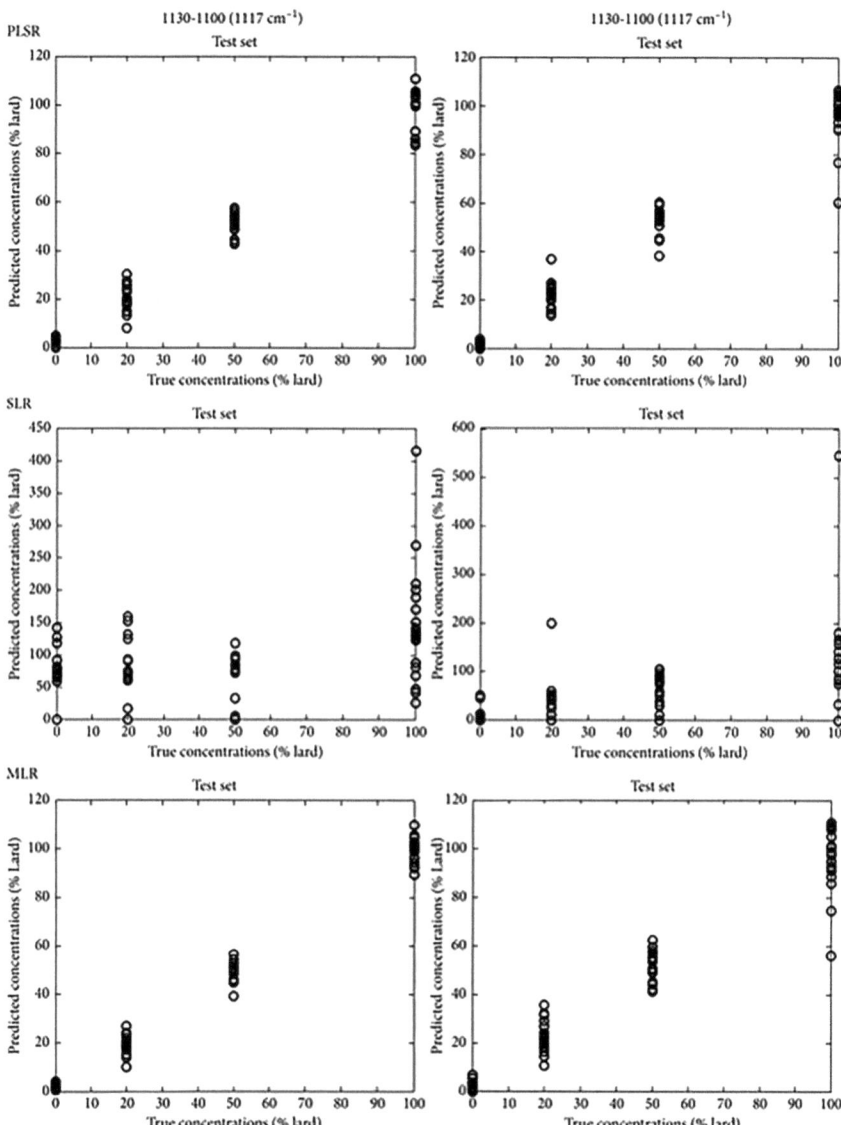

Fig. 4.44 Predicted concentration versus the expected concentration of test samples based on three different models (SLR, MLR, and PLSR) with specific reference to the spectral region of 3006 and 1117 cm^{-1}. (Sim et al. 2018)

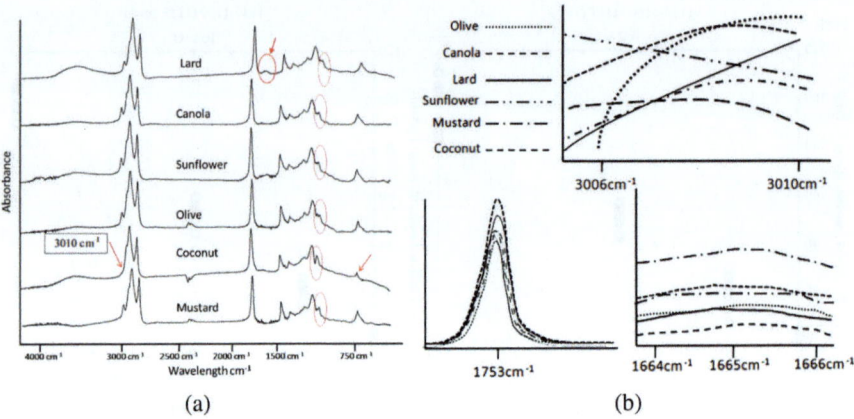

Fig. 4.45 (**a**) Stacked spectra of the edible oils and lard. (**b**) Representative spectra of the oils at different wavelengths. (Munir et al. 2019)

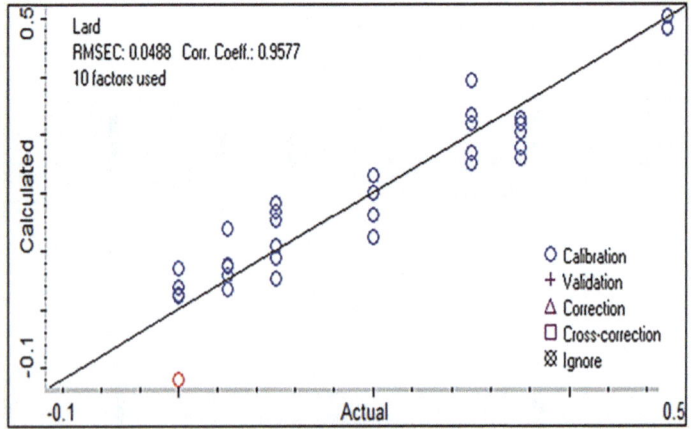

Fig. 4.46 TQ analyst calibration for lard mixing in the region of 1246.75–1078.01 cm^{-1} using PLS. (Munir et al. 2019)

quadrants, facilitating the identification of lard contamination in the imported instant noodle product. In 2019, Munir et al. developed a transmission FT-IR-based technique for the quick identification of lard in mustard, canola, coconut, sunflower, and OO (Fig. 4.45). To achieve this, lard at several concentration ratios (10:0, 9:1, 7:3, 6:4, 4:6, 3:7, 0:10) was added to the selected oils. A single calibration model was developed using a PLS method (Fig. 4.46) for 35 standards, comprising 7 standards from each of the 5 oils, within the frequency range of 1078 to 1246 cm^{-1}. This approach aimed to determine the correlation between the actual adulterant level and the FT-IR predicted levels. The findings indicate that FT-IR with PLS offers a rapid, non-destructive, and efficient approach for detecting lard adulteration in edible oils using a single calibration.

4.7 FTIR Spectroscopy for Halal Authentication

To identify lard adulteration in cooking oil products, Ahda et al. (2024) proposed a methodology based on FTIR spectroscopy and chemometrics. Spectral data were recorded in the 4000–600 cm^{-1} range, with the fingerprint region (1500–900 cm^{-1}) identified as optimal for classification. PCA and PLS-DA were applied, enabling accurate classification of adulterated and pure samples. The quantitative study of lard-CPO, performed at wavenumbers 1500–1085 cm^{-1} and 1800–1600 cm^{-1}, demonstrated a high R^2 of 0.9989 utilizing the calibration eq. $Y = 1.0013x + 0.4712$ and a minimum RMSEC value of 1.001%. The quantitative investigation of lard-VCO at the wavenumber range 1500–1085 cm^{-1} found $R^2 = 0.9994$, employing the eq. $Y = 0.9985x + 0.1352$, with an RMSEC of 0.918%. For the first time, Kuang et al. (2025) proposed a method for rapid detection of horse oil adulteration with sheep oil, butter, and lard by combining IR spectroscopy and deep learning. This study gathered four sample types: horse oil, butter, sheep oil, and lard, which were combined in varying quantities (5%, 10%, 20%, 30%, 40%, 50%) across the wavelength range of 4000–400 cm^{-1}. Additionally, Standard Normal Variable Transformation and Detrending (SNV-DT), along with the incorporation of Gaussian white noise, were applied to obtain 591 × 3601 IR spectral data for each adulteration ratio. An analysis of CNN, RNN, Transformer, and ResNet models commonly employed in the food, cosmetics, and other sectors indicates that fine-tuning ResNet yields the most effective outcomes for detecting applications of contaminated horse oil.

4.7.8 Fish Oil

Fishes are extensively consumed globally and recognized for their significant nutritional benefits for human health. Fish oil is a source of unsaturated fatty acids (monounsaturated and polyunsaturated), which plays a crucial role in nutrition. Additionally, it is abundant in omega-3 fatty acids, including docosahexaenoic acid (DHA) and eicosapentaenoic acid (EPA). The market price of fish oil can be as much as ten times higher than that of other edible oils. Consequently, some unethical producers or distributors can blend it with lower-cost oils to maximize profits (Kwasek et al. 2020). For the Halal authentication of fish oil, FTIR spectroscopy and chemometrics have been reported as an easy and rapid analytical methodology for various oils, including tuna fish oil and cod liver oil (Table 4.12).

FTIR spectroscopy yields distinctive fingerprint spectra (Fig. 4.47a) that are characteristic of Tuna fish oil (TFO), pork oil (PO), and a combination of TFO-PO samples. The resulting FTIR vibrations from the oil and fat samples are mainly from the fatty acids, triglycerides, and triacylglycerols. It could be mentioned that TFO had two (1744 and 1710 cm^{-1}) absorption bands at the carbonyl (C=O) region. The absorption band at 1710 cm^{-1} due to acid is specific for TFO. On the other hand, the absorption band at 1032 cm^{-1} is unique for PO. Therefore, Irnawati et al. (2023) used three types of pattern recognition analysis such as PCA, PLS-DA, and Orthogonal Projections to Latent Structures–Discriminant Analysis (OPLS-DA), to identify the adulterant (PO) with several concentration levels, both absorption bands at 1710 and 1032 cm^{-1} as a potential marker (Fig. 4.47b). It was reported that PCA

Table 4.12 The application of FTIR spectroscopy combined with chemometrics for authentication analysis of fish oils

Fish types/oil	Adulterants	Spectral range	Chemometrics techniques	References
Cod-liver oil	Lard	1500–1030 cm^{-1}	PLS	Rohman and Che Man (2009)
Vegetable and marine oils	Frog fat	2922, 2852, 1745, 1158, and 721 cm^{-1}	PCA	Ali et al. (2015)
Fish oil	Lard, chicken oil, suet, and tallow	4000–550 cm^{-1}	PLS-DA, PCA	Lingzhi et al. (2016)
Fish oil		722 cm^{-1}, 1650–1900 cm^{-1}, and 3006 cm^{-1}	PLS-DA, PLS	Gao et al. (2020)
Tuna fish oil	Pork oil	1710 and 1032 cm^{-1}	OPLS-DA	Irnawati et al. (2023)
Snakehead fish oil		1743, 1654, 1163, and 722 cm^{-1}	PCA, OPLS-DA	Windarsih et al. (2023)
Tuna fish oil		3011, 2922, 1744, and 722 cm^{-1}	SVM	Windarsih et al. (2024)
Cod liver oil	Lard	3009, 2924, 2854, 1743, 1657, 1465, 1377, 1215, 1163, 1117, 1099, and 1032 cm^{-1}	PCA	Zilhadia et al. (2024)

differentiated between TFO and PO samples. However, it could not differentiate authentic TFO from adulterated TFO samples. Therefore, PLS-DA (Fig. 4.48a) and OPLS-DA (Fig. 4.48b) were further used for the discrimination. The S-line plot shown in Fig. 4.49 identified the important variables (11) in predicting the concentration of PO using the OPLS model. Comparatively, the OPLS-DA analysis model successfully detected and predicted the concentration of PO in TFO with high accuracy and good precision, whereas the PLS-DA model only partially discriminated. The proposed method successfully detected PO at a minimum level of 5% in TFO.

Similarly, another study by Windarsih et al. (2024) developed a method for the detection of PO adulteration in TFO using FT-IR coupled with various machine-learning techniques such as SVM, k-Nearest Neighbors (kNN), Artificial Neural Networks (ANN), and Gradient Boosting. FT-IR spectra in the fingerprint region (1400–900 cm^{-1}) were identified as optimal for classification. However, significant spectral differences were observed at 3011, 2922, 1744, and 722 cm^{-1}, corresponding to fatty acid composition variations. Among various machine learning techniques, SVM achieved the highest accuracy ($R^2 = 0.993$, RMSE = 2.719%). The study of the adulteration of cod liver oil (CLO) and emulsion with much cheaper oil-like animal fats (beef, mutton, chicken, and lard) by using FTIR spectroscopy with chemometrics (PLS and PCA) has also been reported by Rohman and Che Man (2009) and Zilhadia et al. (2024). Both studies reported different specific spectral regions for the identification of adulteration of lard in CLO at 1500–1030 cm^{-1}

4.7 FTIR Spectroscopy for Halal Authentication

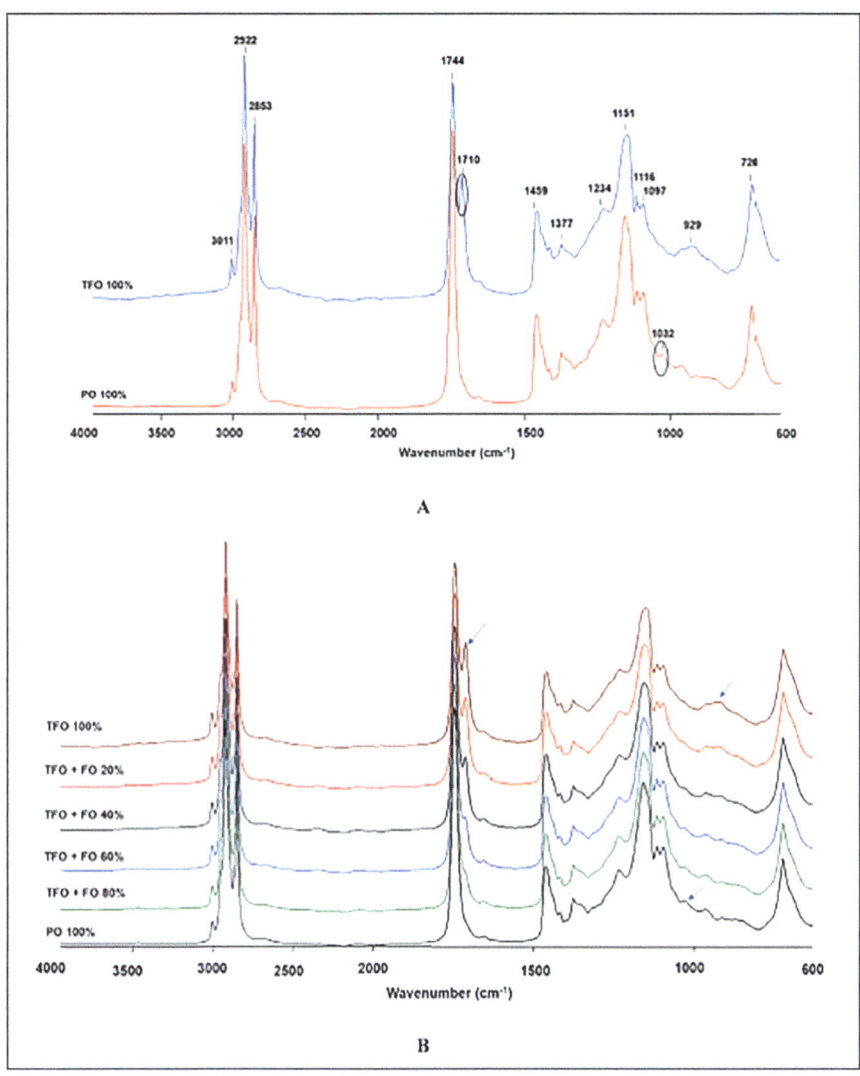

Fig. 4.47 FTIR spectra of TFO and PO (**a**) and TFO adulterated PO (**b**). (Irnawati et al. 2023)

and 1117–1098 cm^{-1}, respectively. Rohman and Che Man (2009) developed a PLS calibration model in the concentration range of (0.5–50% v/v) for quantitative measurement of the adulterant (lard in CLO). A good relationship was reported with an $R^2 > 0.99$ and an RMSECV of 1.04. Zilhadia et al. (2024) applied PCA to the 13 selected points at wavelengths 3009, 2924, 2854, 1743, 1657, 1465, 1377, 1215, 1163, 1117, 1099, and 1032 cm^{-1}, where absorbance was correlated and found 1117 and 1099 specific to identifying the adulteration. The results of PLS and PCA showed that both models successfully predict and differentiate the lard and CLO. The differentiation of frog fat from vegetable and marine oil has been reported

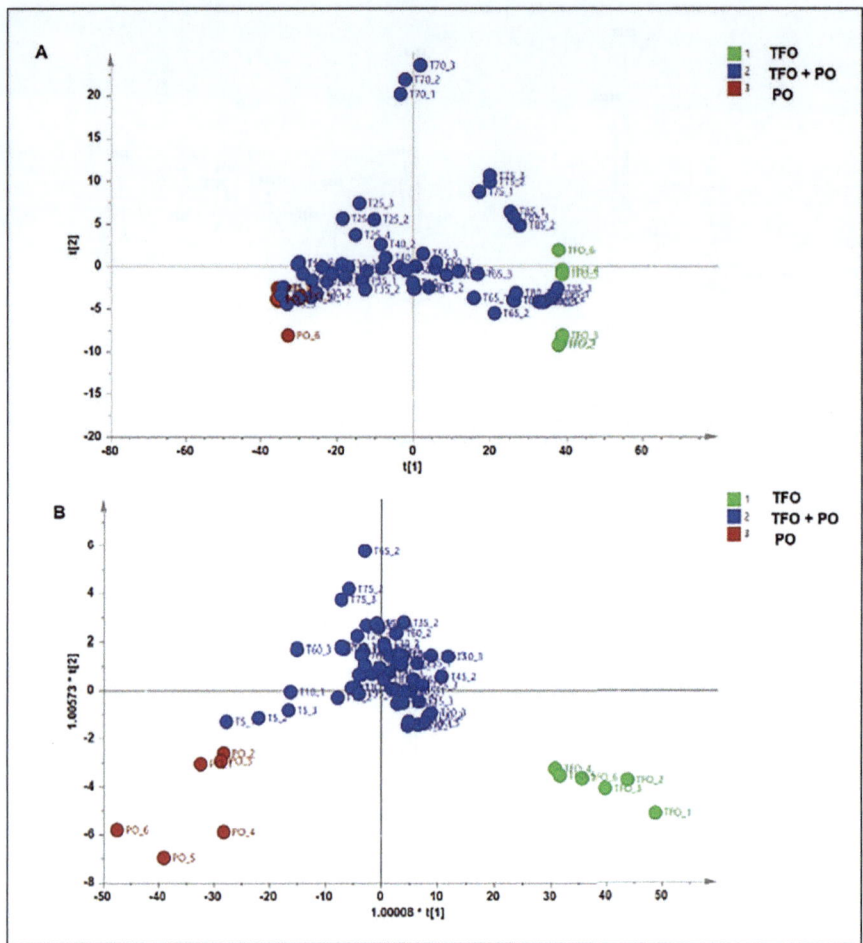

Fig. 4.48 PLS-DA and OPLS-DA for discrimination and classification of authentic and adulterated TFO with PO. (Irnawati et al. 2023)

by Ali et al. (2015) using PCA. The FTIR spectra recorded in the region of 4000–650 cm^{-1} range, identifying 2922 cm^{-1}, 2852 cm^{-1}, 1745 cm^{-1}, 1158 cm^{-1}, and 721 cm^{-1} as key discriminating variables (Fig. 4.50). Frog lipids were effectively categorized by PCA into discrete groups that were separate from vegetable and marine oils, demonstrating definite spectral distinctions. The study demonstrates that FTIR-PCA is a useful method for identifying adulterated frog fat in edible oils, guaranteeing food legality and authenticity.

The DA of different species of terrestrial animal fats and oils in fish oil based on FTIR spectroscopy has been reported by Lingzhi et al. (2016). A total of 27 various species of animal fat (chicken oil, lard, suet, and tallow) and fish oil were evaluated (Fig. 4.51). In the independent validation set, all of the correct discriminant rates

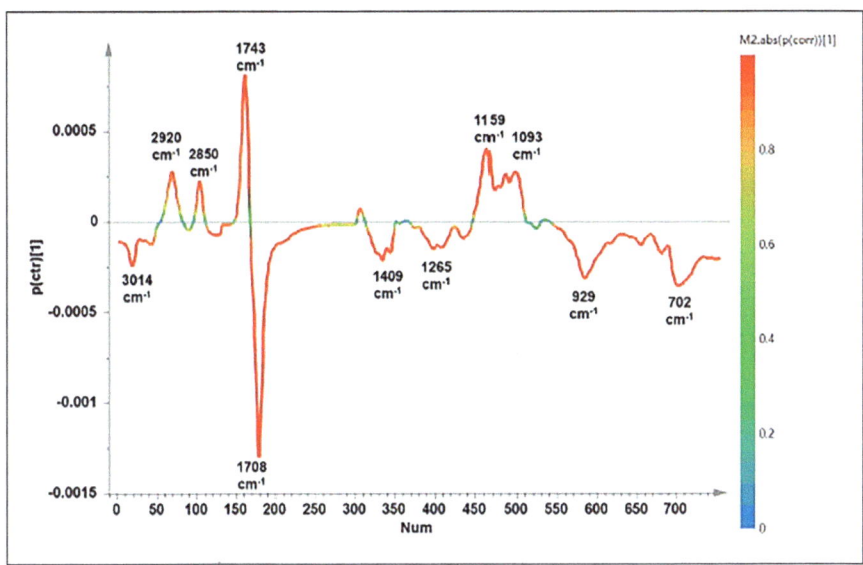

Fig. 4.49 The S-line plot obtained from the OPLS model to identify important variables for predicting the concentration of PO added in TFO. (Irnawati et al. 2023)

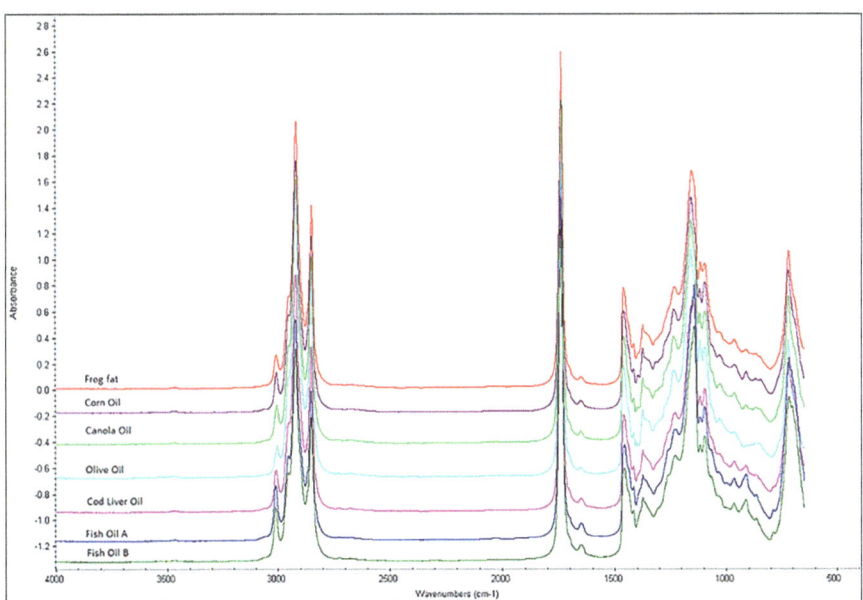

Fig. 4.50 FTIR spectra of frog fats and 3 vegetable oils (corn, canola and olive) and 3 marine 344 oils (cod liver oil, Fish Oil A, Fish Oil B). (Ali et al. 2015)

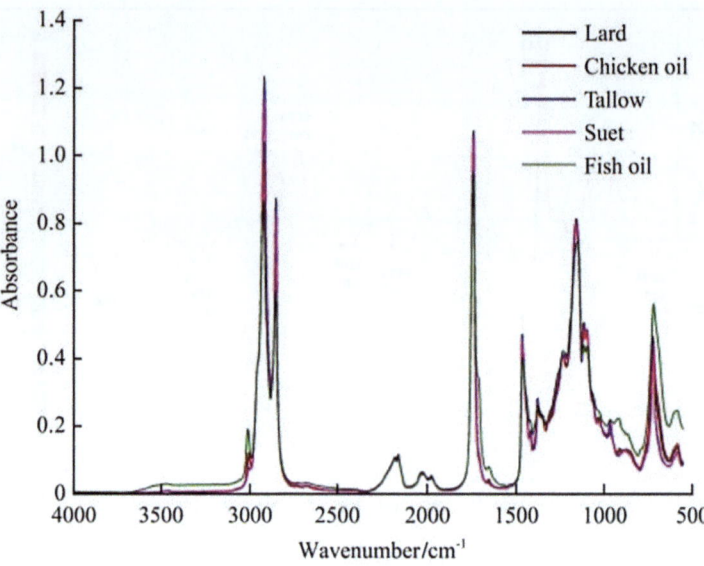

Fig. 4.51 FT-IR spectra of different species of animal fats and oils. (Lingzhi et al. 2016)

were greater than 95% for the adulterated samples, ranging from 1% to 60% (w/w). The best results were obtained when it came to distinguishing the terrestrial fat and oil ingredients (chicken oil, lard, and tallow) in fish oil except suet. Furthermore, the detection limits for the DA of chicken oil, lard, tallow, and suet were 0.6%, 0.8%, 2%, and 3% in fish oil, respectively.

In another study, Gao et al. (2020) developed a method for detecting terrestrial animal lipid adulteration in fish oil using PLS regression and PLS-DA. A total of 736 fish oil samples were analyzed adulterated with lard, chicken oil, tallow, and suet. Spectral data were collected from 4000–550 cm^{-1} with key discriminative bands at 722 cm^{-1}, 1650–1900 cm^{-1}, and 3006 cm^{-1}. The single and multiple adulteration models showed $R^2 > 0.82$ and RPD > 2.5 but were not interchangeable, prompting the development of a combined model with reduced predictive accuracy due to nontarget adulterant interference. Therefore, a sequential strategy was proposed, using PLS-DA for qualitative classification (sensitivity and specificity >0.91), followed by PLS regression for quantitative analysis. For the Halal authentication of snakehead fish oil (SFO), a rapid analytical method has been proposed by Windarsih et al. 2023 using a combination of FTIR spectroscopy and chemometrics. A series of adulterated samples (SFO) in binary mixtures (SFO and PO) with different concentration levels of 0 to 100% v/v were prepared. FTIR spectra of both oils were very similar except for a unique peak at 1418 cm^{-1} in SFO, which is a very small shoulder observed in a PO. PCA, using 12 PCs, successfully differentiated authentic and adulterated SFO samples. The validity of the OPLS-DA model was confirmed by using a permutation test and receiver operating characteristics (ROC) analysis. PLS and OPLS successfully predicted the concentration of PO in SFO with high accuracy and precision (RMSEE and RMSECV) $R^2 > 0.990$ and < 5.0, respectively, in both models. It was elaborated that from the S-line plot, the OPLS model had an important role in the following variables 1708 cm^{-1}, 1153 cm^{-1}, 1107 cm^{-1}, and 914 cm^{-1}.

4.7.9 Gelatin and Gelatin-Based Products

Gelatin and gelatin-based products are widely used in the food, pharmaceutical, and cosmetic industries due to their functional properties, such as gelling, thickening, and stabilizing. Gelatin is a protein obtained from the partial hydrolysis of collagen, primarily sourced from bovine, porcine, or fish origins (Rather et al. 2022). It is widely used in food products such as gummies, marshmallows, and yogurts, as well as in pharmaceuticals (coatings, capsules) and cosmetics. Adulteration of gelatin may involve the use of non-permitted animal sources, synthetic additives, or low-quality substitutes. The authentication of gelatin is critical for Halal, Kosher, and vegetarian dietary requirements. FTIR spectroscopy provides a rapid, non-destructive, and reliable method for verifying gelatin authenticity (Table 4.13).

Hashim et al. (2010) developed a method for discrimination using FTIR with chemometrics. The PCA was employed to classify and characterize gelatin compounds (bovine and porcine) using FTIR spectral regions (3290–3280 cm^{-1} and 1660–1200 cm^{-1}) for calibration models. The findings demonstrate that Cooman's plot clearly distinguished between gelatin samples originating from pigs and cows. The food and pharmaceutical sectors can use the FTIR spectrum to clear up any uncertainty regarding the source of gelatin. Using ATR–FTIR spectra in conjunction with chemometrics, Cebi et al. (2016) developed a fast spectroscopic methodology as an alternate method for the separation and verification of gelatin sources in food items. PCA and HC (Fig. 4.52a, b) were used to clearly distinguish and classify all of the gelatin sources under study (fish, pig, and cow). Chemometric analysis was performed using the Amide-I and Amide-II spectral bands at 1700–1600 cm^{-1} and 1565–1520 cm^{-1}, respectively. Furthermore, pure bovine gelatin was effectively distinguished from a combination of bovine and porcine gelatins using ATR-FTIR spectral data, which is important for the food industry.

In a previous study in 2018, Zilhadia et al. used a combination of ATR-FTIR and PCA to distinguish between cow and pork gelatins and gummy products, including vitamin C. They employed four different types of samples: gelatin from commercial products, cow and pig gelatins produced from simulated vitamin C gummy, and gelatin extracted from commercial substances. Results demonstrate that the PCA plot was scored using the multiple wave numbers in the particular gelatin area, amide A (3296 cm^{-1}), amide I (1633 cm^{-1}), amide II (1552, 1454, 1408, 1338 cm^{-1}), and amide III (1244, 698 cm^{-1}), respectively. Similarly, Cebi et al. (2019) classified and discriminated gelatin gummy candies through FTIR-ATR combined with chemometrics analysis in another study. For chemometric analysis, the spectral region between 1734 and 1528 cm^{-1} was selected. The potential of FTIR spectroscopy to

Table 4.13 The application of FTIR spectroscopy combined with chemometrics for authentication analysis of gelatin, confectionary, and bakery items

Products	Adulterants	Spectral range	Chemometrics techniques	References
Gelatin	Differentiation between porcine gelatin and bovine gelatin	3290–3280 and 1660–1200 cm^{-1}	PCA and DA	Hashim et al. (2010)
Pure bovine gelatin	Bovine and porcine gelatins	1700–1600 and 1565–1520 cm^{-1}	HCA, PCA	Cebi et al. (2016)
Bovine gelatin	Cow and porcine/pork gelatins	3296–698 cm^{-1}	PCA	Zilhadia et al. (2018)
Gelatin in candies	Porcine gelatin	1734–1528 cm^{-1}	Simple regression analysis	Cebi et al. (2019)
Bovine and fish gelatin	Porcine gelatin	1600–1000 cm^{-1}	PCA and LDA	Hassan et al. (2021)
Bovine/soft candy	Porcine gelatin	4000 to 500 cm^{-1}	PLS	Salamah et al. (2023)
Gummy candy	Pork	1181 and 2940 cm^{-1}	PCA, PLS	Supandi et al. (2024)
Bovine gelatin	Porcine gelatin in marshmallows	1697–1093 cm^{-1}	PCA	Fatmawati et al. (2024)
Chocolate	Lard, cocoa butter	4000–650 cm^{-1}	PLS	Che Man et al. (2005)
Cake lipids	Lard	1117–1097, 990–950 cm^{-1}		Syahariza et al. (2005)
Chocolates, cake, and biscuits		1500–800 cm^{-1}, 1780–1700 cm^{-1}, and 3500–2900 cm^{-1}		Syahariza (2006)
Biscuit lipids		3050–2800, 1800–1600, and 1500–650 cm^{-1}	PLS-DA	Che Man et al. (2011a, b)
Puff pastry		1159 cm^{-1}, 1115 cm^{-1}, and 1096 cm^{-1}	DA and PLS	Rani et al. (2009)
Imported chocolate products		4000–650 cm^{-1}	PCA and PLS	Suparman et al. (2015)
Chocolate		4000–400 cm^{-1}	PLS	Bahri and Che Man (2016)
Bread formulation		1190–900 cm^{-1}	Simple regression analysis	De Cicco et al. (2019)
Chocolates and biscuits	Lard and palm oil	3035–2984 cm^{-1}	PLS	Kunbhar et al. (2025)

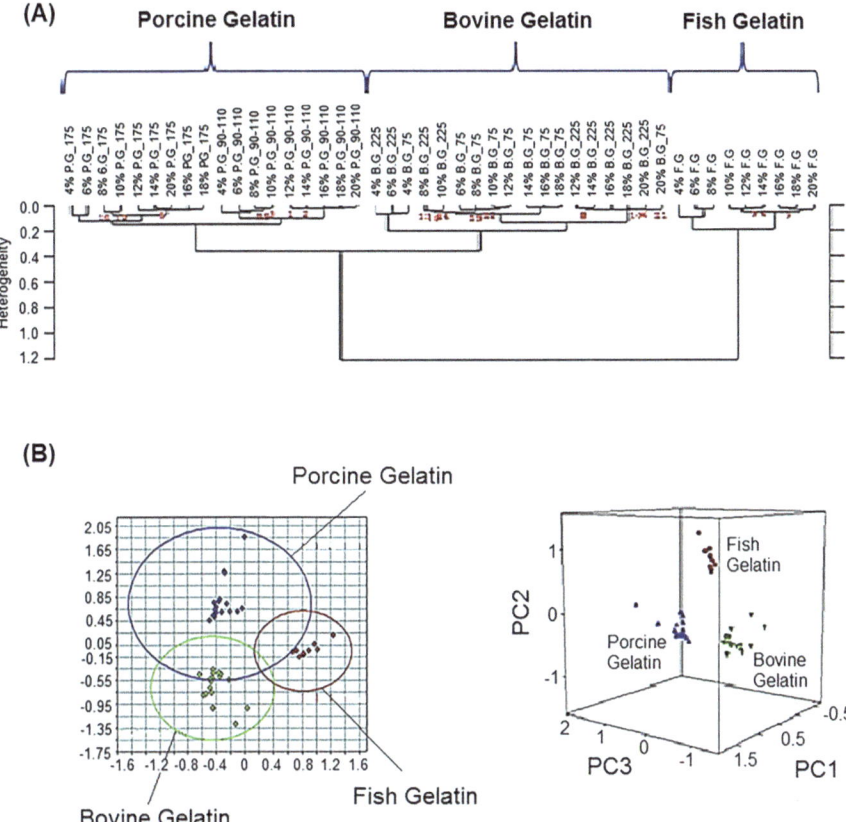

Fig. 4.52 (**a**) Dendrogram of cluster analysis (Ward's algorithm), (**b**) two and three-dimensional PCA analysis map of FT-IR spectra from the total of 45 different standard samples of bovine, porcine, and fish gelatins (**b**) (P.G: Porcine gelatin; B.G: Bovine gelatin). (Cebi et al. 2016)

identify bovine and porcine sources in gummy candies was evaluated and validated using real-time PCR. Gummy candies were successfully classified and differentiated based on the gelatin source (porcine and bovine) with 100% accuracy using the ATR technique without requiring any sample preparation. Hassan et al. (2021) discriminate porcine, bovine, and fish gelatin by using a chemometrics fuzzy autocatalytic set (c-FACS) in conjunction with FTIR spectroscopy. PCA and LDA were used to compare the gelatin spectra at the Amide band in the 1600–1000 cm^{-1} areas using c-FACS (Figs. 4.53 and 4.54). According to the c-FACS technique results, each pig, bovine, and fish gelatin had distinct signatures with prominent wavenumbers of 1444–1450 cm^{-1}, 1470–1475 cm^{-1}, and 1496–1500 cm^{-1}, correspondingly (Figs. 4.55 and 4.56). The c-FACS method was further performed on published data on meat products (chicken and pork) to validate the technique (Fig. 4.57). For this, the data on chicken and pork meats were adopted from the study of Al-Jowder et al. (1997). The c-FACS method was rigorous and quicker than PCA and LDA in differentiating the gelatin sources.

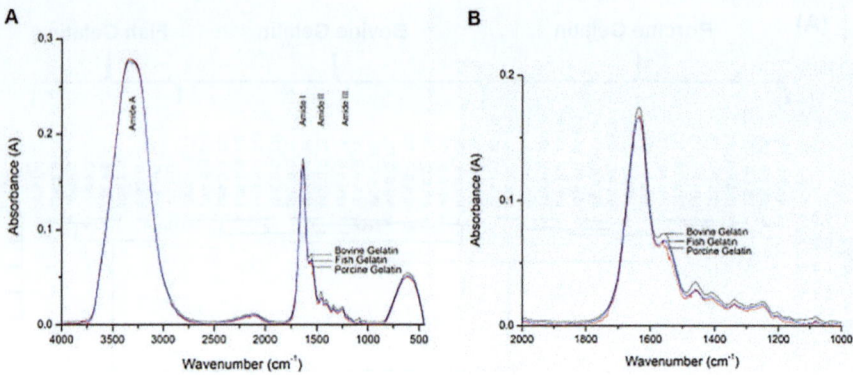

Fig. 4.53 (**a**) Full-scale FTIR spectra of bovine, porcine, and fish gelatins and (**b**) enlarged spectra of bovine, porcine, and fish gelatins at 2000–1000 cm^{-1}. (Hassan et al. 2021)

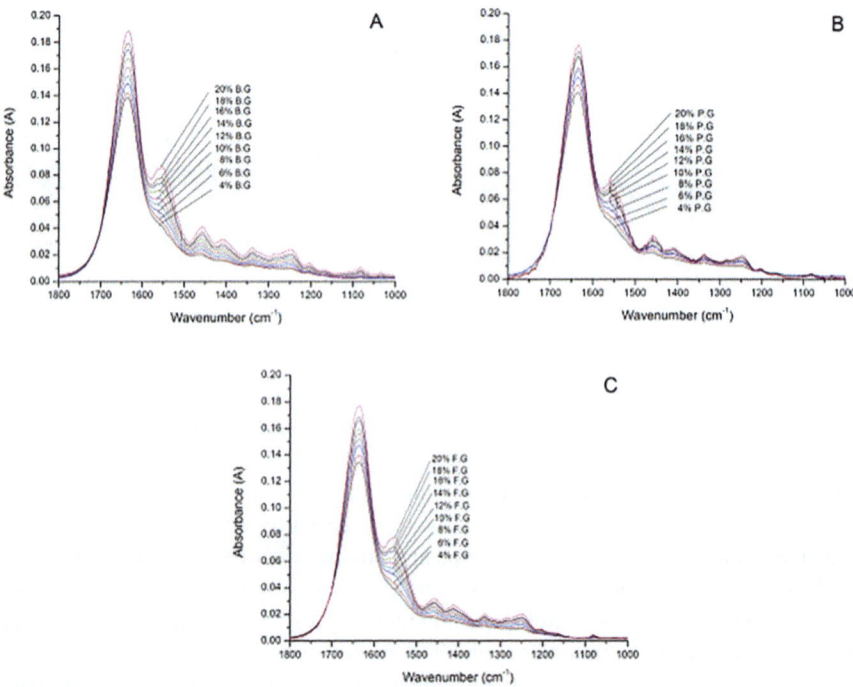

Fig. 4.54 Concentration-dependent FTIR spectra of (**a**) bovine gelatin (B.G), (**b**) porcine gelatin (P.G) and (**c**) fish gelatin (F.G). (Hassan et al. 2021)

4.7 FTIR Spectroscopy for Halal Authentication

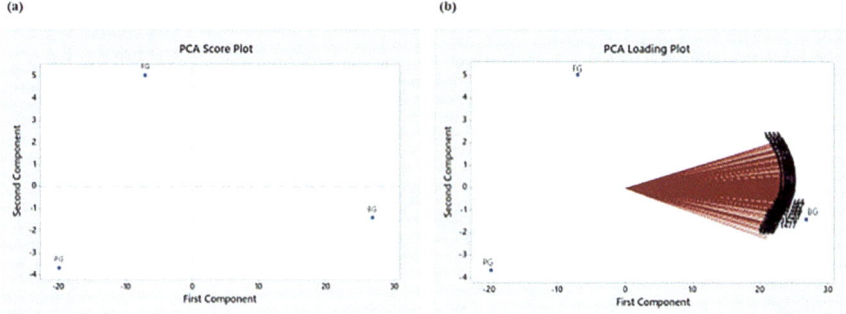

Fig. 4.55 PCA (**a**) score plot and (**b**) loading plot for bovine (BG), porcine (PG), and fish (FG) gelatins. (Hassan et al. 2021)

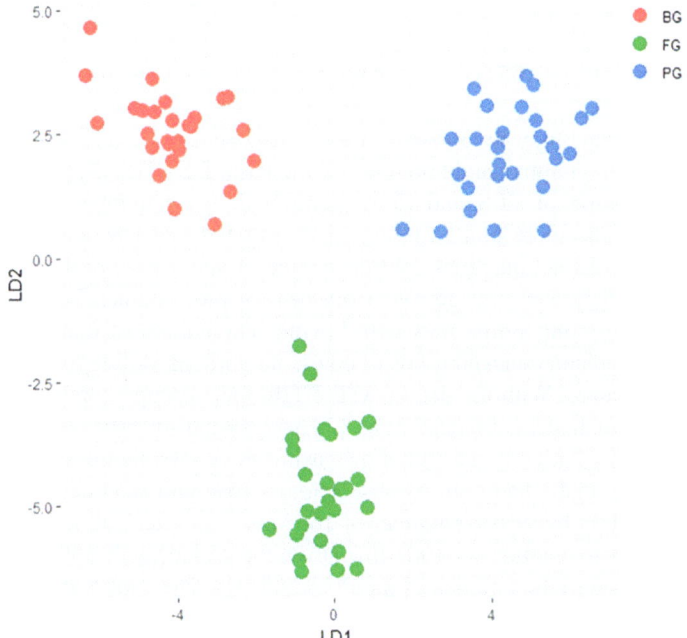

Fig. 4.56 LDA plot for bovine (BG), porcine (PG), and fish (FG) gelatins. (Hassan et al. 2021)

Salamah et al. (2023) analyzed the porcine and bovine gelatins in soft candy using FTIR and chemometrics. Porcine gelatin formulations with concentrations of 15, 30, 45, 60, 75, 90, and 100% were used to make the reference candy samples. FTIR was used to measure all of the candy samples in the reflection mode, with wavenumbers ranging from 4000 to 500 cm^{-1}. PLS calibration findings for y = 0, 99,999x + 0.000396. The R^2 value is 0.99999 RMSECV = 3.69% for internal validation and R2 = 0.9994, RMSEP = 1.28% for external validation and RMSEC is

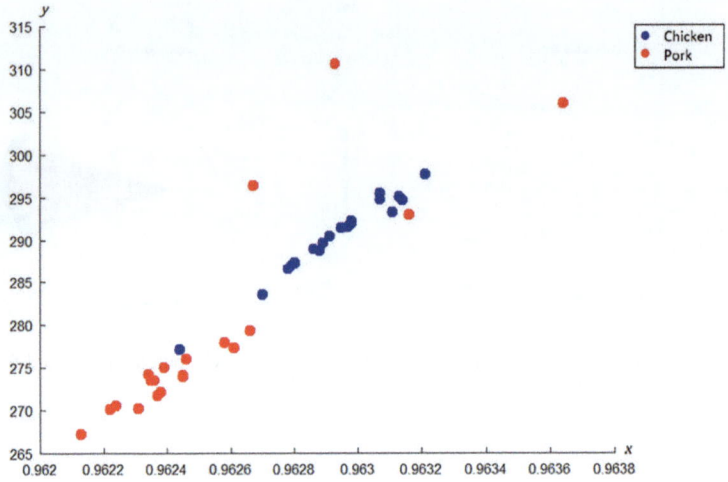

Fig. 4.57 Coordinated c-FACS of chicken and pork meats. (Hassan et al. 2021)

0.03%. In a recent study by Supandi et al. (2024) aimed to study the gelatin and pork contained in gummy candy products in the Pasar Baru Bekasi, FTIR was used in the combination of PLS and PCA along with wave numbers 1181.57 and 2940.87 cm^{-1}. The score plot's findings demonstrated that cow and pork gelatin were located in distinct quadrants. While samples 1 and 2 were in their quadrants, indicating that they were not derived from beef or pig, samples 3 and 4 were in quadrant 3, suggesting similarities with bovine. The pattern of linear absorbance varies with the gelatin concentrations of cows and pork, according to the multivariate regression curve. With the use of ATR-FTIR spectroscopy combined with the chemometrics methods, Fatmawati et al. (2024) distinguished the presence of porcine and bovine gelatins in marshmallows using PCA. The findings demonstrated that the porcine gelatin had a strong absorption at 1697 and 1654 cm^{-1}, and bovine gelatin at a slightly lower wavenumber of 1638 cm^{-1}. In selected regions, 100% of bovine gelatin marshmallows (S100) differ from those made with a blend of porcine gelatin, according to the findings of the PCA analysis. The pattern of linear absorbance varies with the amounts of porcine gelatin, according to the multivariate regression curve. The wavenumber 1093 cm^{-1} has the largest coefficient equation (Fig. 4.58).

4.7.10 Confectionary and Bakery Products

Confectionery and bakery products, including chocolates, pastries, candies, and bread, are widely consumed across the globe. These products are composed of a variety of ingredients, including fats, sugars, emulsifiers, additives, and flavoring agents. The industry faces challenges related to product adulteration, such as the use of non-food-grade ingredients, synthetic sweeteners, and the substitution of

4.7 FTIR Spectroscopy for Halal Authentication

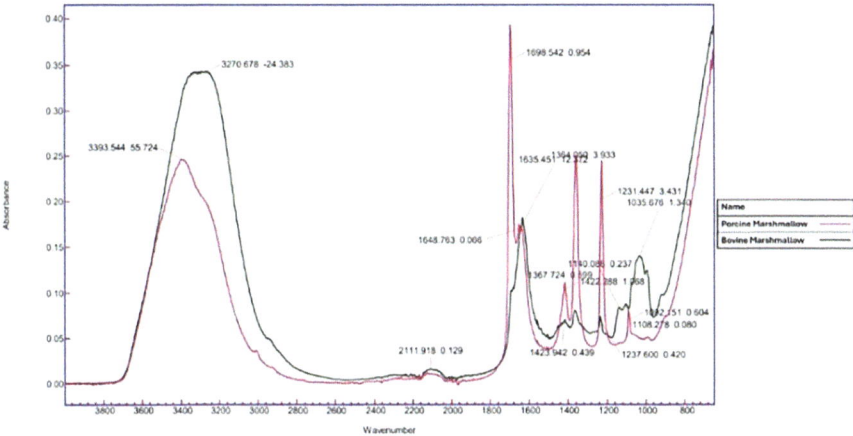

Fig. 4.58 Overlay FTIR spectrum of porcine and bovine gelatin extracted from marshmallow. (Fatmawati et al. 2024)

premium ingredients with cheaper alternatives. Authenticating confectionery and bakery products is vital for consumer safety and brand reputation. FTIR spectroscopy offers a rapid, non-destructive, and reliable method for verifying product Halal authenticity (Table 4.13).

The first study was conducted by Che Man et al. (2005) to detect the presence of lard in the chocolate formulation using FTIR spectroscopy with chemometrics. Following the recording, interpretation, and identification of the bands in the spectrum corresponding to lard, cocoa butter, and their mixture (comprising 0% to 15% lard in cocoa butter), a semi-quantitative method was suggested to determine the percentage of lard in blends based on spectral data at the entire (4000–650 cm^{-1}) frequency region of MIR using the equation y = 0.9225x + 0.5539, with a standard error (SE) of 1.305 and the coefficient of determination (R2) was 0.9872 (Fig. 4.59).

Syahariza et al. (2005) used FTIR spectroscopy to quickly ascertain the amount of lard in cake composition. The FTIR spectroscopic calibration model was derived using a PLS chemometric method in the selected regions 1117–1097 and 990–950 cm^{-1}. By contrasting the FTIR spectroscopy results with the actual value, the models' coefficient of determination (R^2) was calculated. The obtained R^2 was 0.9790, while the calibration SE was 1.7520. Similarly, in 2006, Syahariza et al. detected lard in chocolate, cake, and biscuit formulations by using FTIR spectroscopy combined with ATR and PLS regression. The results demonstrate the in biscuits, FTIR analysis across 1500–800 cm^{-1}, 1780–1700 cm^{-1}, and 3500–2900 cm^{-1}, resulted in a strong linear correlation (R^2 = 0.9974, SEC = 2.819), confirming the accuracy of FTIR in predicting lard content across different food matrices. While in chocolate, a high correlation (R^2 = 0.9872, SE = 1.305) was achieved using spectral data from 4000–650 cm^{-1}. Furthermore, for the cake, lard detection was conducted by incorporating it into shortening (0–100%), with PLS regression models developed in the 1117–1097 cm^{-1} and 990–950 cm^{-1} regions, yielding R^2 = 0.9937

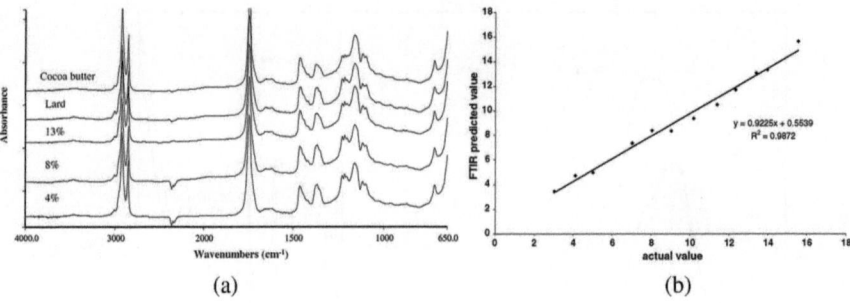

Fig. 4.59 (a) FTIR spectra of lard, cocoa butter, and some blends of lard–cocoa butter. (b) Data from actual value versus FTIR-predicted values for lard content by the "leave-one-out" cross-validation

(SE = 2.257) in cross-validation and improved SE (1.752) in test validation. Rani et al. (2009) differentiated the spectral data of 21 puff pastry samples using FTIR spectroscopy coupled with ATR accessory prepared from three types of fats individually (butter, fat, and margarine) and admixture of each. In this study, the qualitative and quantitative data of the fat profiles were examined by using the combination of DA and PLS at 1159 cm^{-1}, 1115 cm^{-1} and 1096 cm^{-1} regions. PLS coupled with DA detected as low as 0.5% content of lard in the binary mixture of butter-lard and margarine-lard. Che Man et al. (2011a, b), in another study, conducted an analysis and discrimination of specific edible fats and oils utilized in biscuit preparation through the application of FTIR spectroscopy with chemometric technique. To classify lard and other commercial vegetable oils and animal fats, DA was used with specific frequency zones (3050–2800, 1800–1600, and 1500–650 cm^{-1}). The analysis revealed that all vegetable fats and oils, along with animal fats such as lard, formed a clearly defined cluster in the Coman plot. Additionally, real food, specifically biscuits, was divided into two groups using DA. Suparman et al. (2015) conducted FTIR and GCMS spectrometry to authenticate Halal status across various imported chocolate products available in the market. The analysis of FTIR spectra for lard and chocolate within the wave number range of 4000–650 cm^{-1} reveals distinct characteristics specific to lard at the wavenumber of 3006 cm^{-1}, 1118 cm^{-1}, and 1097 cm^{-1}. The examination of calibration models (PCA and PLS) within the fingerprint region 999–1190 cm^{-1} serves as a method for identifying lard in chocolate fat. The correlation between the actual and predicted values of lard in chocolate is represented by the eq. Y = 1.000x-0.0378, with R^2 value of 0.997 and RMSEC 1.563, indicating a minimum detection limit at a concentration of 4%.

Bahri and Che Man (2016) studied and discovered the spectral bands associated with lard, cocoa butter, and a combination of the two (ranging from 0 to 10% lard in cocoa butter). Using FTIR spectral data from the band within the frequency range 4000–400 cm^{-1} and applying the equation y = 1.0144x–0.0644, a semi-quantitative method was presented to determine the percentage of lard in the chocolate formulation. The R^2 value was 0.9892, and the corresponding standard error was 0.4504. De Cicco et al. (2019) developed a multi-analytical approach for detecting pig-derived

4.7 FTIR Spectroscopy for Halal Authentication

ingredients in bread, addressing concerns about undeclared animal-origin components in baked products through ATR-FTIR spectroscopy, gas chromatography, PCR, and HPLC. The results demonstrate that the ATR-FTIR effectively discriminated lard-containing (1% w/w) from conventional bread with frequency regions 1190–900 cm^{-1}, while GC analysis provided characteristic fatty acid profiles for lard-containing samples. The study highlights PCR and proteomics as precise tools for detecting porcine enzymes, whereas GC and ATR-FTIR are valuable for identifying lard, offering robust screening methods for food authentication and regulatory enforcement. In order to determine the amount of lard in imported chocolates and biscuits, Kunbhar et al. (2025), in a recent study, developed an analytical method based on ATR-FTIR with chemometrics (Fig. 4.60 and Fig. 4.61). A calibration graph of lard oil (2–35%) with palm oil in the 3035–2984 cm^{-1} region was created, which showed a correlation coefficient (R^2) = 0.9994 and minimum errors in RMSEC and RMSEP of 0.320 and 0.315, respectively (Fig. 4.62). The accuracy of the calibration model was assessed by using the RMSECV, which yielded 1.17 with the best LOD and LOQ 0.10% and 0.35%, respectively.

4.7.11 Dairy Products

Dairy products are essential components of human nutrition, providing a rich source of vitamins, fats, proteins, and minerals. Ensuring the authenticity and quality of dairy products is crucial to prevent adulteration and maintain consumer trust (Shawky et al. 2024). FTIR spectroscopy is an effective analytical tool for authenticating dairy products, including milk fat, butter, pure ghee, and cheese. There are many published studies on the adulteration of dairy products using FTIR spectroscopy. These studies highlight various adulteration techniques, the effectiveness of FTIR in detecting them, and the development of spectral databases for authentication (Table 4.14 and Fig. 4.63).

4.7.11.1 Milk Fat

Milk fat (MF) is perceived as one of the highly nutritious fat products because of its functional properties. MF is rich in nutrients, including lipid soluble vitamins, particularly vitamin D, and various fatty acids that contribute positively to health. It is common practice to adulterate with inferior fats in order to increase the benefits (Mohan et al. 2021).

Detection and quantification of lard oil (LO) in binary mixtures with Bovine milk fat (BMF) by using FTIR spectroscopy with chemometrics has been reported by Windarsih and Irnawati (2020). The results demonstrated that the BMF and adulterated BMF with LO were perfectly identified by DA utilizing FTIR spectra (normal) at the wavenumber of 3098–669 cm^{-1} (Fig. 4.64). Furthermore, for the measurement of LO in BMF, PLS calibration of FTIR spectra using 1st derivative at

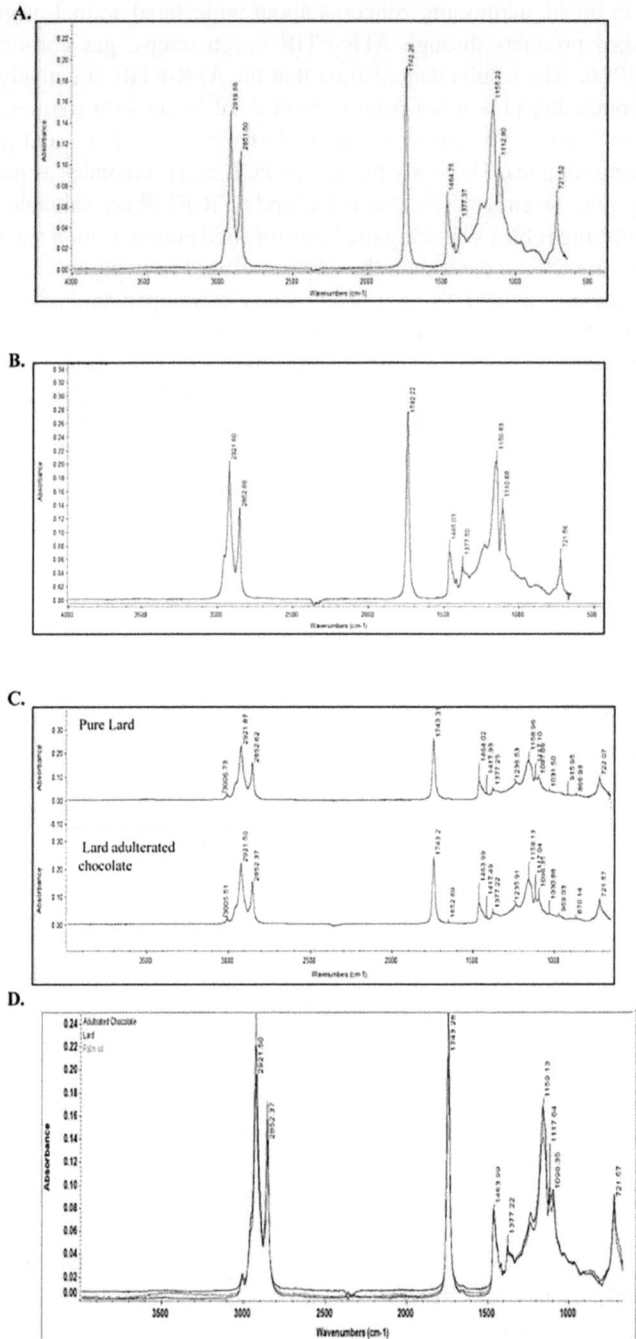

Fig. 4.60 Representative FTIR-ATR spectra of pure (**a**) chocolate fat, (**b**) biscuits fat, (**c**) pure lard in comparison to chocolate adulterated with lard, and (**d**) pure lard and Pure palm in comparison to adulterated chocolate. (Kunbhar et al. 2025)

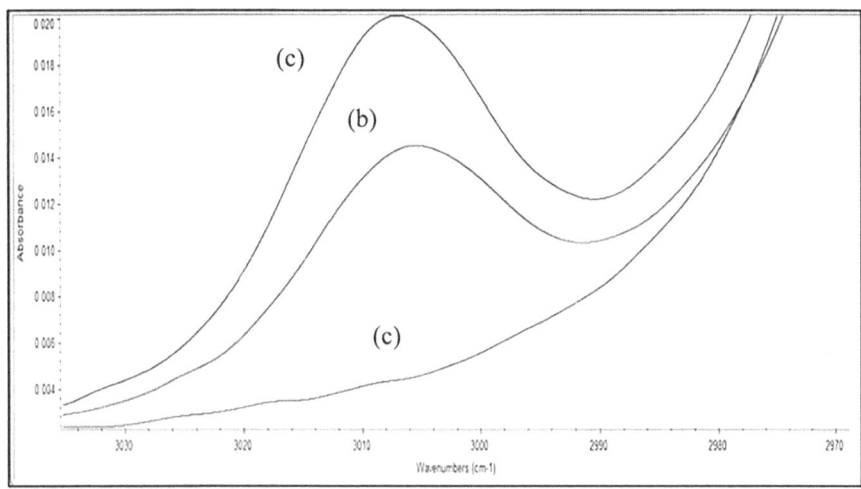

Fig. 4.61 FTIR spectra of (**a**) biscuit, (**b**) pure lard oil, and (**c**) chocolate adulterated with lard at region 3035–2984 cm^{-1}. (Kunbhar et al. 2025)

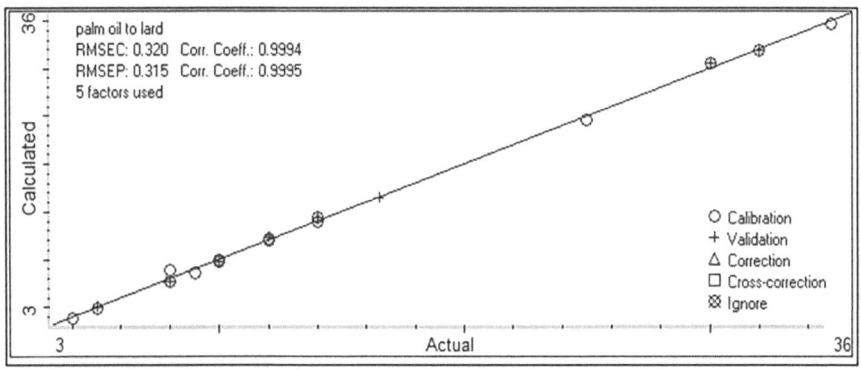

Fig. 4.62 Partial least square (PLS) calibration model of the actual value of lard in (% v/v) versus FTIR predicted value (%). (Kunbhar et al. 2025)

the combination of wavenumbers 3033–2770 cm^{-1} and 1510–692 cm^{-1} was used. This method had a high R^2 value (>0.99) in both calibration and validation models, as well as reduced RMSEC (0.631) and RMSEP (1.94).

Similarly, Arifah et al. (2022) used FTIR spectroscopy in combination with chemometrics for authentication of various milk samples, including horse milk (HM) from goat milk (GM) and cow milk (CM). ATR spectral measurements were performed directly on milk in the mid-IR (4000–650 cm^{-1}) range. The strongest model for the link between the actual values of GM and the FTIR projected values was produced using PLSR utilizing 2nd derivative spectra at 3200–2800 and 1300–1000 cm^{-1}, as opposed to PCR. RMSEC and RMSEP values were 0.0093 and

Table 4.14 The application of FTIR spectroscopy combined with chemometrics for authentication analysis of dairy and cosmetics products

Dairy/cosmetics	Products	Adulterated products	Spectral range	Spectral treatment and chemometrics	References
Milk fat	Bovine milk fat	Lard oil	3098–669 cm^{-1}	PLS	Windarsih and Irnawati (2020)
	Goat's and cow's milk	Horse's milk	4000–650 cm^{-1}	PLS	Arifah et al. (2022)
	Cow milk	Donkey milk	1000–2500 nm	PCA and PLS-DA	Di Donato et al. (2023)
Butter	Butter	Lard	3005–725 cm^{-1}	PLS	Nurrulhidayah et al. (2015)
		Margarine and pork fat	904 to 1699 nm	PCA and PLSR	Kazazić et al. (2021)
Ghee	Ghee	Pig body fat	4000 and 500 cm^{-1}	PCA	Upadhyay et al. (2018)
Cheese	Cheese samples	Lard	700, 1140–1070, 756, and 720 cm^{-1}	PLS	Alkhalf et al. (2017)
Cosmetics	Cream		3020–2995 and 1200–1000 cm^{-1}	PLS-DA	Rohman and Che Man (2011a, b)
	Lotion		1200–1000 cm^{-1}	PLS	Lukitaningsih et al. (2012)
	Cream		1785–702, 3020–2808, and 1200–1000 cm^{-1}	PCA	Rohman et al. (2014)
	Lipstick		1200–800 cm^{-1}	PLS and PCA	Waskitho et al. (2016)
	Soap formulation		1453 and 1415 cm^{-1}	PCA	Aziz et al. (2023)

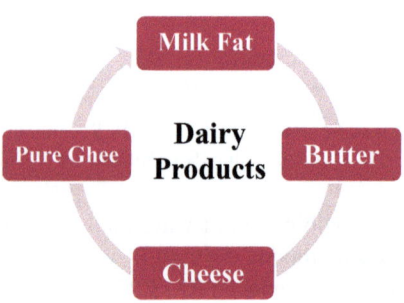

Fig. 4.63 Some common dairy products

0.0794, respectively, while the R^2 values were under these conditions for the calibration (0.9995) and validation (0.9612) models. The authentication of HM from CM demonstrated minimal errors with an RMSEC of 0.0164 and RMSEP of 0.0336. Utilizing PLSR with normal FTIR spectra in the ranges of 3800–3000 and 1500–1000 cm^{-1} yielded an R^2 value exceeding 0.99 for the correlation between actual values of CM and the FTIR projected values in both calibration and validation models.

4.7 FTIR Spectroscopy for Halal Authentication 133

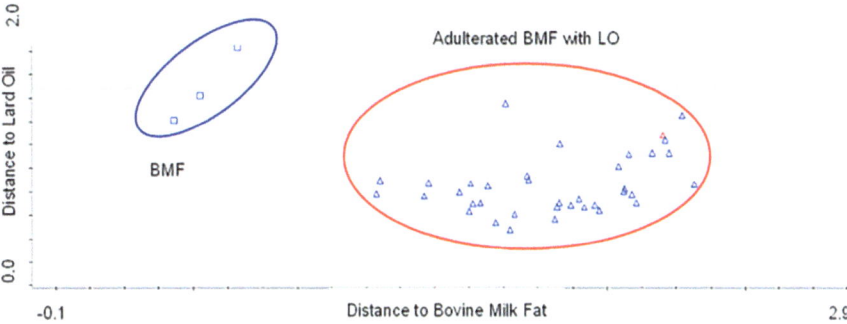

Fig. 4.64 Discriminant analysis of authentic bovine milk fat (BMF) and adulterated BMF with lard oil (LO) in the wavenumber of 1510–692 cm^{-1}

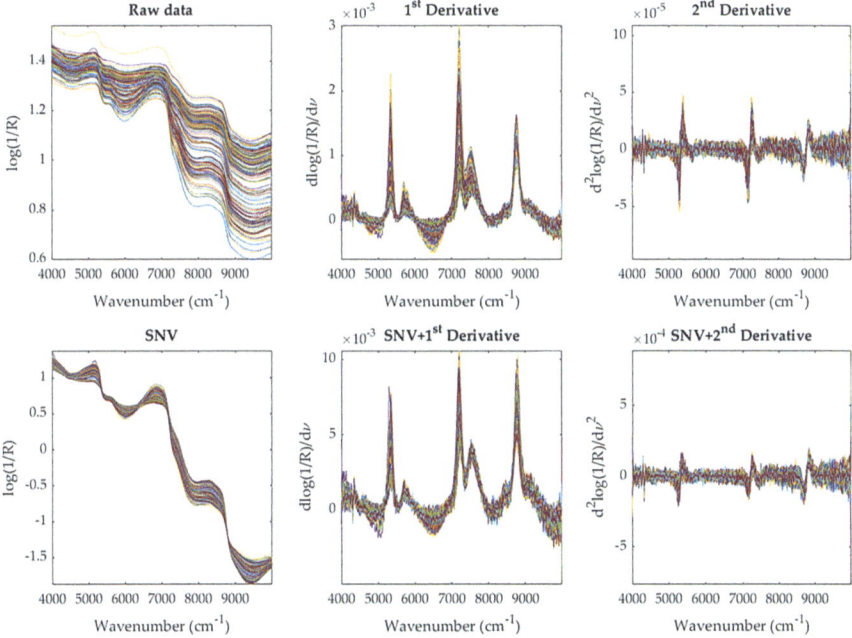

Fig. 4.65 Raw and differently pre-processed (1st derivative, 2nd derivative, SNV, SNV+1st derivative, SNV+2nd derivative) NIR spectra collected on the milk samples. (Di Donato et al. 2023)

The use of NIRS (Fig. 4.65) with chemometric methods to verify donkey milk (DM) adulterated with CM was studied by Di Donato et al. (2023). Advanced chemometric models, including PCA and PLS-DA (Figs. 4.66 and 4.67), were used to categorize milk samples and identify adulteration based on the NIR spectra, which were obtained in the 1000–2500 nm range. Based on the results, NIRS, supported by chemometrics, offered a dependable and non-invasive method to differentiate DM from other kinds of milk accurately.

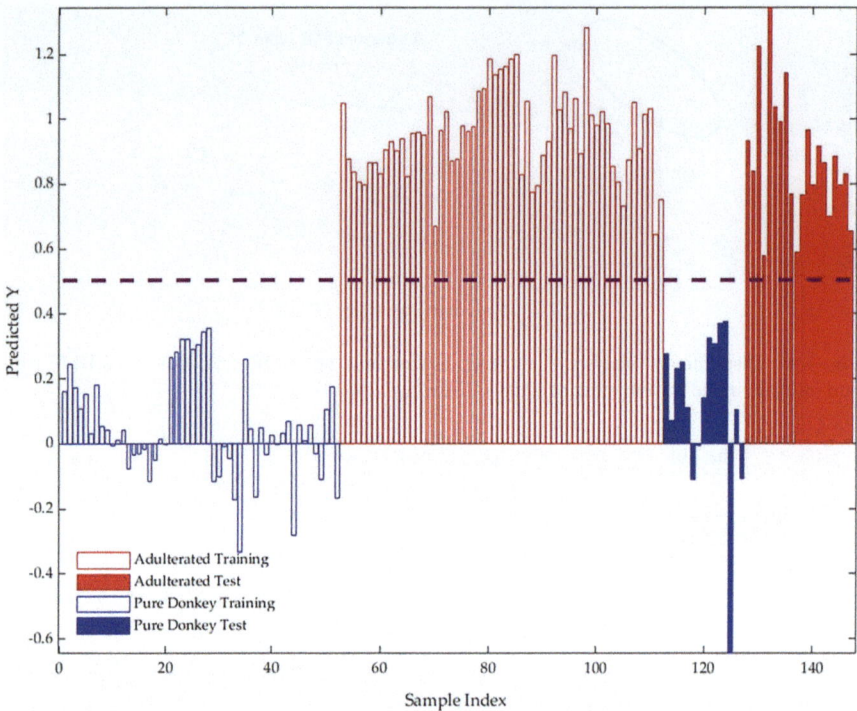

Fig. 4.66 PLS-DA analysis: values of the predicted response calculated on the training (empty bars) and test (full bars) samples. The violet dashed line is the classification threshold. (Di Donato et al. 2023)

4.7.11.2 Butter

Butter is a concentrated form of milk fat obtained through churning cream. It is an essential ingredient in various culinary applications and a significant source of dairy fat. However, butter is susceptible to adulteration with vegetable oils, synthetic additives, and other non-dairy fats, which can impact its quality and nutritional value (Panchal and Bhandari 2020). Authenticating butter composition is vital for regulatory compliance and consumer protection. FTIR spectroscopy provides a rapid, non-destructive, and reliable method for verifying butter authenticity.

In 2015, Nurrulhidayah et al. performed a quantitative analysis of butter containing lard through the application of ATR-FTIR and a chemometric approach. The findings indicated a distinction between non-mixed butter and lard-containing butter in the wave numbers ranging from 3005 to 725 cm^{-1}. The analysis employed a PLS regression to predict the amount of lard. The model achieved an impressive regression R^2 with low RMSEE and RMSEP, 0.999, 0.0947, and 0.0687, respectively. The findings suggest that ATR-FTIR coupled with PLS techniques are suitable for analyzing lard in butter. In a 2021 study, Kazazić et al. compared NIR and GC as rapid tools for detecting butter adulteration with margarine and pork fat. PCA and PLSR were used to categorize and measure the amounts of adulteration in butter samples with known adulterant concentrations. NIR spectra were collected

4.7 FTIR Spectroscopy for Halal Authentication

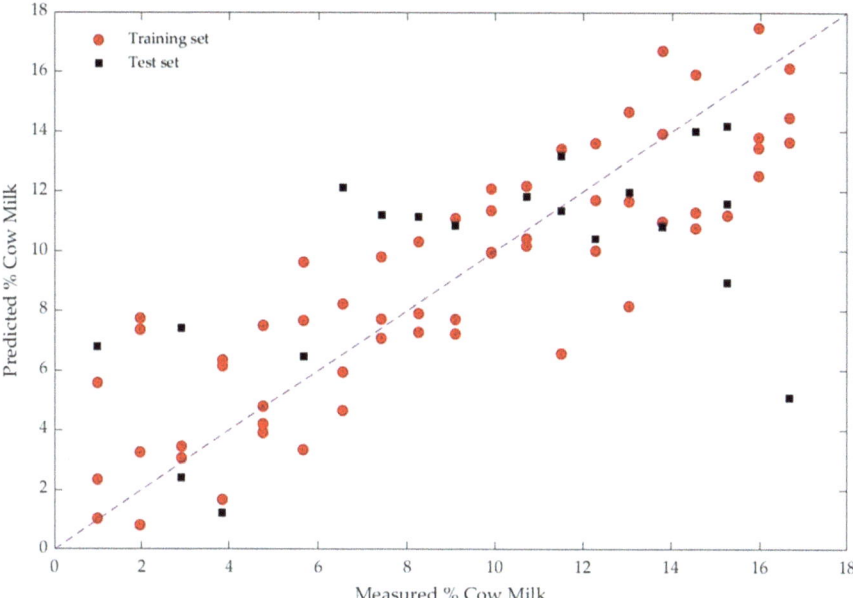

Fig. 4.67 PLS regression. Predicted percentage of cow milk in the different mixtures with donkey milk. (Di Donato et al. 2023)

within the 904–1699 nm wavelength range. The samples were analyzed qualitatively using PCA, and quantitative predictions were made using the PLSR approach. For the first time, this study shows that samples tainted with pork fat and margarine can be effectively separated and that amounts of added pork fat and margarine can be successfully determined using NIR.

4.7.11.3 Ghee

Ghee, a clarified form of butter, is a widely consumed dairy product valued for its rich aroma, nutritional benefits, and extended shelf life. Ghee is obtained by heating butter or cream to remove water and milk solids, leaving behind pure butterfat (Kumar et al. 2018). It is widely used in cooking, traditional medicine, and religious rituals. However, ghee is often adulterated with vegetable oils, animal fats, or synthetic additives, affecting its quality and authenticity. Authentication of ghee is essential for regulatory compliance, consumer protection, and health safety. FTIR spectroscopy provides a rapid, non-destructive, and reliable method for verifying ghee authenticity.

According to Upadhyay et al. (2018), the presence of pig body fat in pure ghee can be detected using chemometrics and ATR-FTIR spectroscopy (Fig. 4.68). The spectra of the spiked and pure (ghee and pig body fat) samples were acquired in the MIR region between 4000 and 500 cm^{-1}. PCA on the selected wavenumber range was used to create different clusters of the data at the 5% level of significance (Fig. 4.69). The prediction of probable class membership was conducted using the

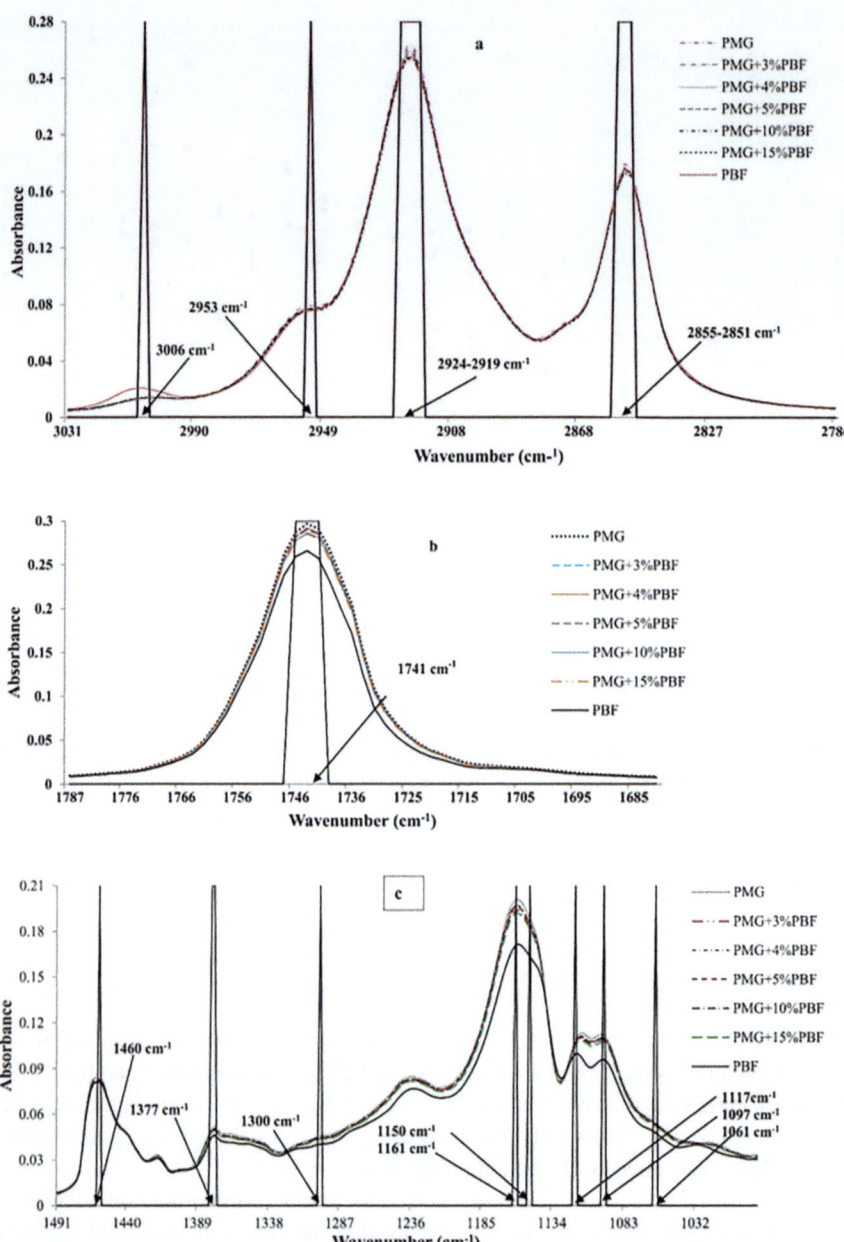

Fig. 4.68 Spectra of pure mixed ghee along with the addition of pig body fat in the wavenumber range (**a**) 3030–2785 cm^{-1} (**b**)1786–1680 cm^{-1} (**c**)1490–919 cm^{-1}. PMG—pure mixed ghee (1:1—pure cow ghee and pure buffalo ghee) PBF—pig body fat. (Upadhyay et al. 2018)

4.7 FTIR Spectroscopy for Halal Authentication

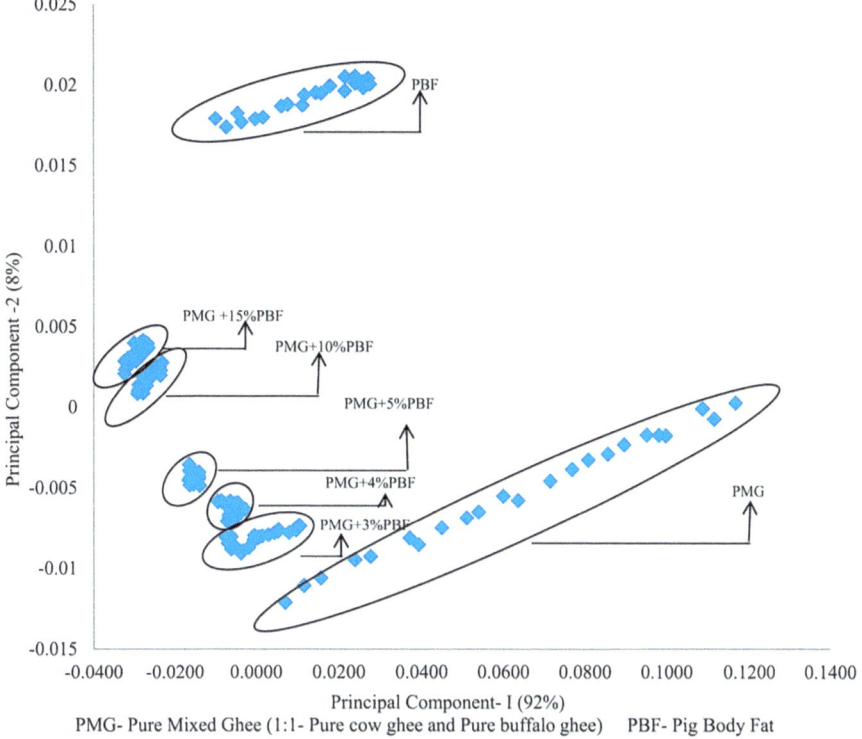

Fig. 4.69 Principal component scores plot depicting clusters of pure mixed ghee, pure pig body fat, and mixed ghee with different levels of pig body fat in the wavenumber range of 3030—2785 cm^{-1}. (Upadhyay et al. 2018)

SIMCA approach. Pig body fat and pure ghee never misclassified themselves, and almost 90% of the samples were correctly classified into their appropriate classes. Using the PLS approach, the R^2 value for both the calibration and validation sets was more than 0.99. The study found that pig body fat might spike in pure ghee as little as 3%, respectively.

4.7.11.4 Cheese

Cheese is produced by coagulating milk proteins, primarily casein, through enzymatic or acid-based processes. The wide variety of cheese products arises from differences in milk source, bacterial cultures, aging conditions, and processing methods. However, cheese is often adulterated with non-dairy fats, protein extenders, or synthetic additives, which affect its nutritional and sensory qualities (Li et al. 2023). FTIR spectroscopy is an effective analytical tool for authenticating cheese products, detecting adulteration, and verifying compositional integrity. Alkhalf et al. (2017) carried out a study to distinguish three different types of cheese lipids from lard using FTIR and other methods of analysis. Their findings indicated that GC-MS, FTIR spectroscopy, and DSC are reliable analytical techniques for qualitatively differentiating lard from other fats. The FTIR spectra showed characteristic absorption bands at 3007, 1140–1070, 756, and 720 cm^{-1}, which correspond to specific molecular vibrations.

4.7.12 Cosmetics

The Halal aspects of cosmetics have gained significant attention due to the inclusion of both haram and Halal ingredients in personal care and cosmetic products. To ensure purity, quality, safety, and compliance with Halal requirements, all ingredients used in Halal cosmetics must be verified. Rules for the production, labeling, and sale of cosmetics have been developed by regulatory agencies, including the USFDA (2004), the EU Cosmetic Directive (1976), and the ASEAN Cosmetic Directive (2008), with a focus on consumer safety. This safety standard also applies to Halal cosmetics, guaranteeing that these items are devoid of harmful substances. In the cosmetic industry, Halal certification implies that products do not contain porcine by-products, alcohol, or derivatives. Detecting haram and Halal sources in cosmetic products is critical and can be achieved through various techniques (Khattak 2009). FTIR Spectroscopy has emerged as a reliable method for identifying and authenticating ingredients. The examination of adulteration of oils and fats with more affordable alternatives, such as animal fats, has gained prominence and frequency in recent years. The technique uses specific infrared fingerprint regions to differentiate between components, ensuring the authenticity of ingredients and compliance with Halal standards (Table 4.14).

The first study was conducted by Rohman and Che Man 2011a, b), which focused on the quantification and classification of lard in cream formulations through the application of FT-IR spectroscopy combined with chemometrics. PLS and DA were employed across two frequency ranges, i.e., 3020–2995 cm^{-1} and 1200–1000 cm^{-1}. The PLS calibration model for the relationship between the actual (x-axis) and FT-IR predicted (y-axis) values of lard demonstrate a coefficient of determination (R) of 0.997 and anRMSE of 2 calibrations (RMSEC) of 0.808% (v/v), resulting in the equation y = 0.997x + 0.065. In another study, Lukitaningsih et al. (2012) utilized FTIR spectroscopy in conjunction with chemometrics and identified lard in a binary mixture with palm oil in a lotion formulation. A similar frequency range of 1200–1000 cm^{-1}, as reported earlier by Rohman and Che Man (2011a, b), was employed to measure the amount of lard in the lotion formulation's mixture with palm oil. With a coefficient of determination (R2) of 0.99, the PLS calibration model showed a strong connection between the FTIR projected value and the actual value of lard. The levels of lard in the mixture with EVOO in cosmetic cream have also been analyzed by Rohman et al. (2014) using FTIR spectroscopy in conjunction with PCA utilizing the integrated frequency ranges of 1785–702 cm^{-1} and

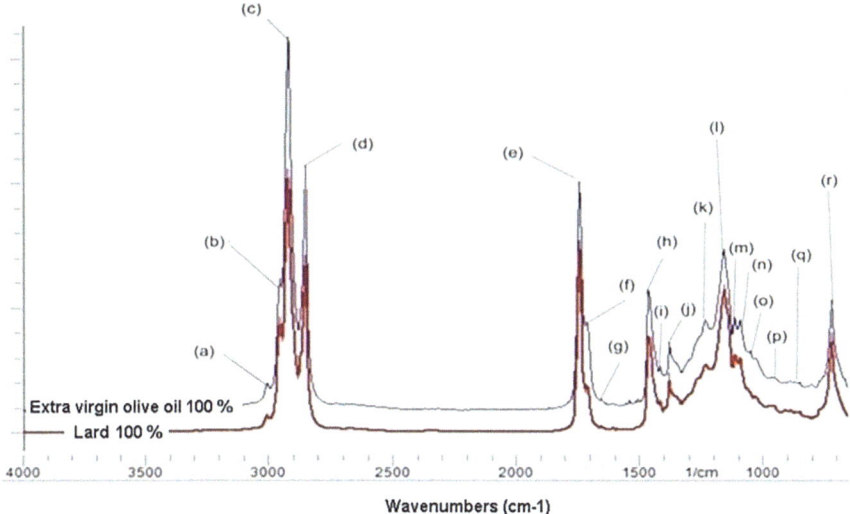

Fig. 4.70 FTIR spectra of lard and extra virgin olive oil obtained from lipid extraction of cream containing 100% lard or 100% EVOO. (Rohman et al. 2014)

3020–2808 cm^{-1} (Fig. 4.70). PCA has been successfully used to classify creams with and without lard in the formulation using absorbance intensities at 1200–1000 cm^{-1} as variables. The RMSEP value was 3.61%, and the R^2 value was 0.991, respectively, for the lard-predicted values.

Similarly, Waskitho et al. (2016) identified lard extracted from lipstick formulation containing castor oil using FTIR spectroscopy combined with PLS and PCA. Lard was analyzed both qualitatively and quantitatively using PCA and PLS analysis (Figs. 4.71 and 4.72). Lard was successfully identified and quantified using PCA and PLS in the wavelength range of 1200–800 cm^{-1} using the Bligh & Dyer as the extraction technique yielded the best results. Furthermore, the model (PLS) showed the lowest RMSEP and the highest determination coefficient (R^2).

Moreover, a recent study by Aziz et al. (2023) used FT-IR and multivariate analysis to determine and measure the presence of lard as an adulterant in a binary mixture with palm oil within cosmetic soap formulations. The FTIR spectrum of palm oil and lard differed slightly at the wavenumber range of 1453 and 1415 cm^{-1} in palm oil and lard, respectively. Two PCA models were developed (Model A and Model B), which detailed a comparable cumulative variability (CV) of 92.86% for the complete dataset. Moreover, the lowest mean square error (MSE) and relative standard error (RSE), as well as the highest determination coefficient (R^2) of 0.996 by MLR and PCR, suggest that both regression models effectively quantified the lard adulterant present in cosmetic soap.

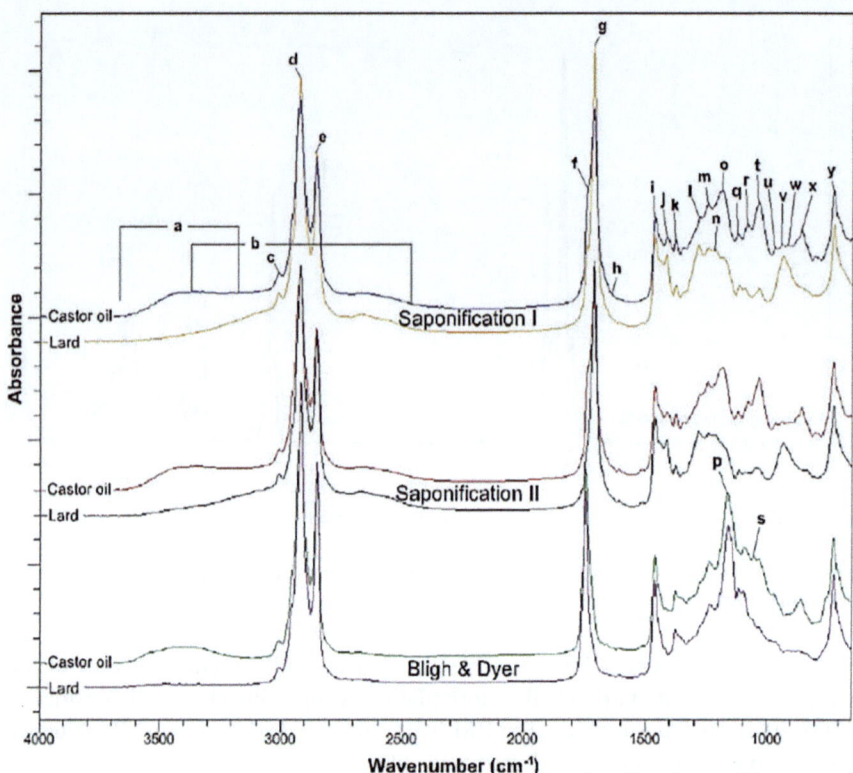

Fig. 4.71 The FTIR spectra of lard and castor oil extracted by three extraction methods from prepared lipstick formulation. (Waskitho et al. 2016)

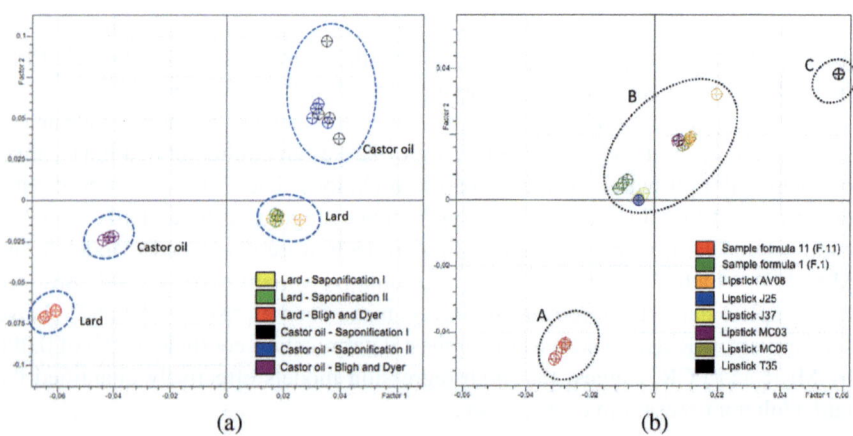

Fig. 4.72 (a) Score plot of lard and castor oil extracted by three extraction methods (saponification method I, saponificationmethod II, and Bligh and Dyer) from lipstick formulation. (b) Score plot of lard and castor oil extracted by Bligh & Dyer method from commercial lipsticks. (Waskitho et al. 2016)

4.7.13 Beverages and Drinks/Alcohol

Alcohols are widely used across various industries due to their versatile chemical properties. In Islam, alcohol is strictly prohibited due to its harmful effects on individuals and society. The primary reason for this prohibition is the impairment of judgment and self-control, which can lead to sinful behavior and societal harm (Dama et al. 2025). Furthermore, the consumption, production, and trade of alcohol are also forbidden, emphasizing a holistic approach to its avoidance. The role of alcohol in Islam, therefore, is one of strict prohibition, ensuring spiritual, physical, and social well-being (Sheikh and Islam 2018). FTIR has been widely utilized in the analysis of beverages, with most research concentrated on alcoholic beverages and the determination of a wide variety of parameters (Uríčková and Sádecká 2015). The following are a few studies reflecting the use of FTIR spectroscopy to authenticate alcohol content.

Gallignani et al. (2005) introduced an analytical approach for quantifying ethanol content in alcoholic beverages and wine (Fig. 4.73). The idea was based on a direct FTIR detection of the analyte in the organic phase after a quantitative online liquid-liquid extraction of ethanol with chloroform. A baseline was set between 844 and 929 cm^{-1} to adjust the quantification based on the band at 877 cm^{-1}. The results for ethanol determination in wine, as well as in other beverages (alcoholic) such as beer and spirits, were satisfactory.

Lachenmeier et al. (2005) and Lachenmeier (2007) employed FTIR spectroscopy combined with chemometrics (PLS and PCA) for the quality control and authenticity assessment of alcoholic beverages, including tequila, beer, and spirit drinks. The first study focused on authentic Mexican tequilas (14) and commercial samples (24) in the spectral region of 4996–930 cm^{-1}. From mixed tequilas, PCA distinguished 100% agave tequilas with higher methanol and isobutanol content (Figs. 4.74 and 4.75). The second study extended the first approach to spirit drinks

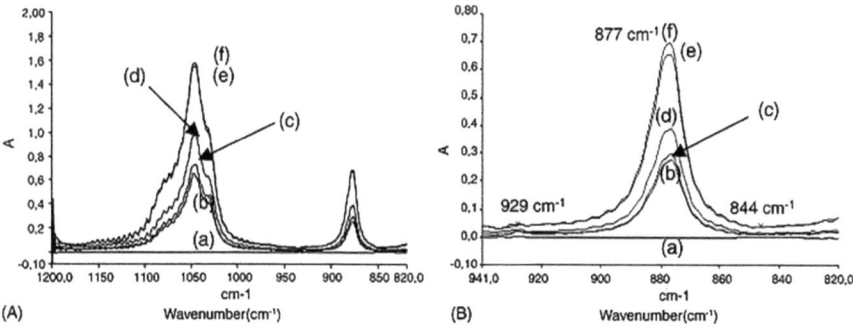

Fig. 4.73 FTIR spectra of different commercial alcoholic beverages. (A) Amplification of the 1046 cm^{-1} band. (B) Amplification of the 877 cm^{-1} band. Samples: (a) DI, (b) beer (4.5%, v/v), (c) aqueous standard of ethanol (5.0%, v/v), (d) red wine (12.5%, v/v) diluted 50% (v/v) with DI, (e) white rum (40%, v/v) diluted 25% (v/v) with DI, (f) whisky (43%, v/v) diluted 25% (v/v) with DI. (Gallignani et al. 2005)

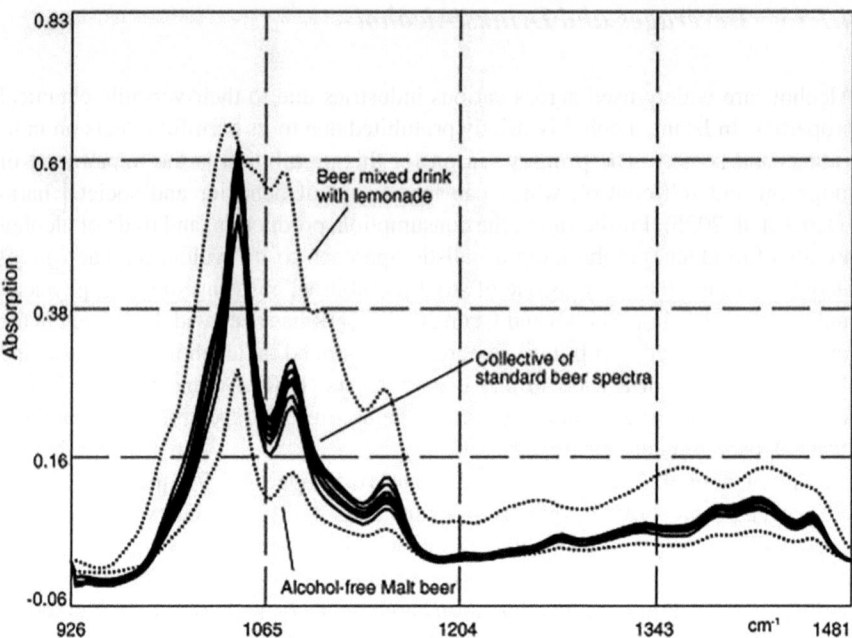

Fig. 4.74 Spectral approach for outlier detection of beer samples in the characteristic range between 926 and 1481 cm^{-1}. (Lachenmeier 2007)

(including vodka, rum, brandy, whiskey, and absinthe) and beers (Wheat beers, Pilsener, Lager, and Bock beers) analyzed in the region of 926–5012 cm^{-1}. PLS models employed which showed strong correlations (R^2 = 0.97–0.98) for various spirit and beer parameters.

Similarly, Coldea et al. (2013) reported a quantification method for the determination of methanol and ethanol in fruit brandies from various Transylvania-Maramureş regions by using FTIR spectroscopy combined with chemometrics. Results demonstrated that the 1200–950 cm^{-1} spectral range revealed methanol peaks at 1020 and 1112 cm^{-1} and ethanol peaks at 1047 and 1087 cm^{-1}. PCA analysis in the 1170–1000 cm^{-1} range successfully differentiated brandy samples based on their biological and geographical origin. Three different studies were reported by Debebe et al. (2017a, 2017b, 2017c) for the development of fast, accurate, and nondestructive chemometric-based NIR and MIR methodology for alcohol authentication (ethanol and methanol content) in distilled alcoholic beverages. For the NIR and MIR analyses, 1660–1720 nm, 1200–850 cm^{-1}, and 1180–950 cm^{-1} ranges were used. MIR confirmed methanol presence/absence, while NIR independently measured alcohol strength without sample preparation. The model, validated with varying concentrations of standards (2–15% ethanol, 0.1–1% methanol), achieved high accuracy (R^2 = 0.999 for ethanol, R^2 = 0.929 for methanol) and recovery rates of 97–98%. The methods showed a strong correlation with GC. All three studies highlight that NIR and MIR spectroscopy are fast, accurate, and non-destructive

4.7 FTIR Spectroscopy for Halal Authentication

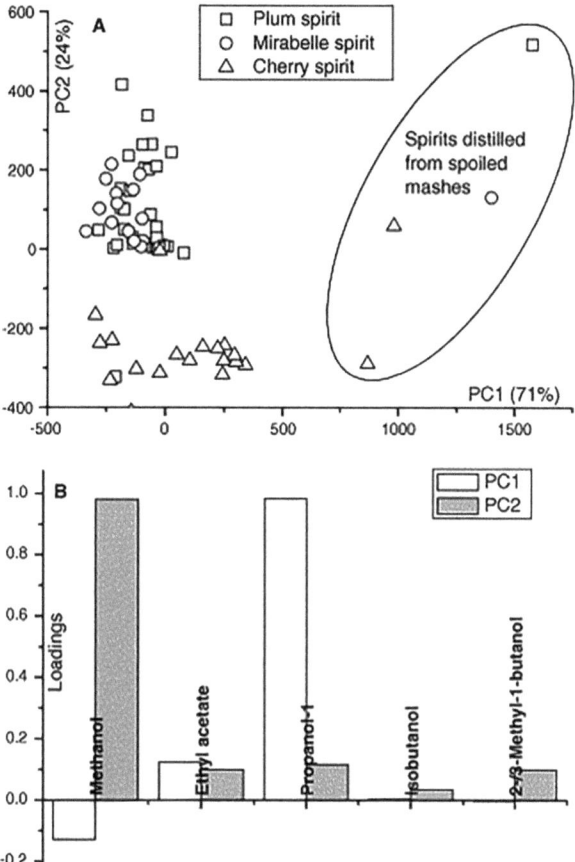

Fig. 4.75 PCA of the quantitative results of fruit spirits. (**a**) PCA scores plot and (**b**) corresponding loadings. (Lachenmeier 2007)

tools for alcohol authentication. In another study, da Costa Fulgêncio et al. (2023), in recent research, employed a multi-product model by combining PLS and portable NIRS for the determination of ethanol content in fermented alcoholic beverages. NIR spectra were obtained from 908 and 1676 nm for alcoholic beverage samples (Fig. 4.76), reflecting an alcohol concentration range of 4.3 to 15.3% (v/v). The PLS model yielded excellent outcomes, with RMSEC and RMSEP at 0.8% and 0.9%, respectively (Fig. 4.77). The simultaneous quantification of methanol and ethanol in different Portuguese wines was investigated by Thanasi et al. (2024) utilizing the FTIR and PLS regression. Both the solutions presented characteristic vibration frequencies at 1047 cm^{-1} and 1087 cm^{-1} for ethanol and 1020 cm^{-1} and 1112 cm^{-1} for methanol. The findings revealed a significant relationship with the reference values for the ethanol determination (y = 0.9557x, R^2 = 0.988). However, the model demonstrated insufficient predictive ability regarding the methanol content.

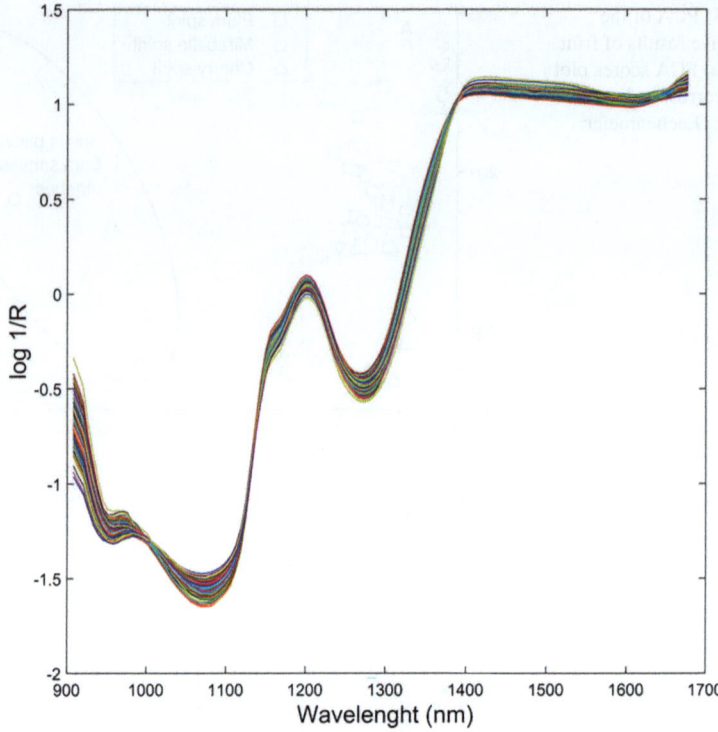

Fig. 4.76 Preprocessed NIR spectra of analyzed beverage samples. (da Costa Fulgêncio et al. 2023)

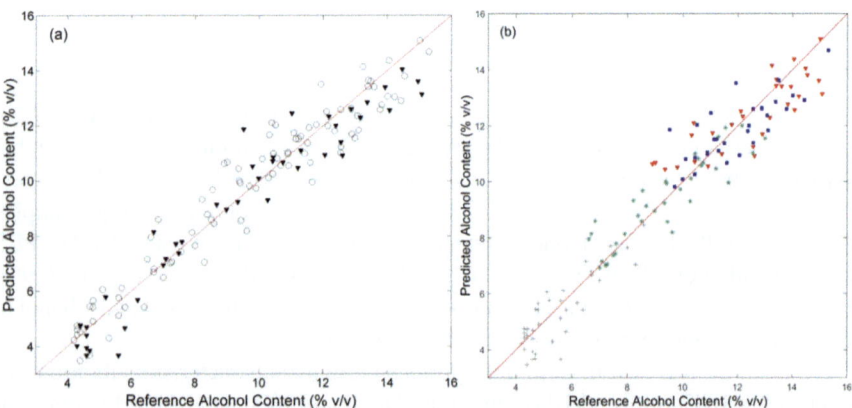

Fig. 4.77 Plot of reference alcohol content versus predicted values for the developed PLS model, (**a**) representation as a function of calibration and validation samples and (**b**) type of beverage. Gray crosses represent beer samples, green asterisks cider samples, blue squares mead samples, and red down triangles wine samples. (da Costa Fulgêncio et al. 2023)

4.8 Portable and Hand-Held FTIR Spectroscopy to Food Control

When real-time data collection in the field is required, hand-held and portable versions of FTIR spectrometers can be employed. The food sector now has a new avenue for on-site and real-time assessment of food quality and production process thanks to a miniature version of vibrational spectroscopy equipment. The portable instruments (FTIR) are built through the integration of compact interferometers and circuits within the system's framework (McVey et al. 2021). Previous studies indicate that portable and hand-held MIR spectrometers may encounter certain limitations associated with essential system components, including moving parts and quantum-type pyroelectric detectors, as these components necessitate cooling in order to reduce the level of thermal noise (Rodriguez-Saona et al. 2020). Nonetheless, earlier studies have indicated that MIR spectroscopy offers distinct advantages over NIRS, as it delivers unique insights into the functional groups of organic materials and allows for the quantification of individual compounds based on their absorption intensities (Cebi et al. 2023). Multiple studies have been conducted to investigate the feasibility of employing portable spectrometers for the purpose of evaluating the quality of food products (Halal authentication), as shown in Table 4.15.

Schmutzler et al. (2015) developed three NIR-based multivariate analytical methods for detecting pork and pork fat adulteration in veal sausages on-site testing (handheld spectrometer) within the 6028–5480 cm^{-1} spectral range (Fig. 4.78).

Table 4.15 Overview of commercially available portable spectroscopy devices for the determination of food authenticity

Meat/animal fat/oil products	Adulterants products	Device	Spectral range	Chemometric analysis	References
Veal product	Pork	Thermo Fisher Scientific microPHAZIR	6028–5480 cm^{-1}	PCA, SVM	Schmutzler et al. (2015)
Palm oil	Lard	Viavi MicroNIR™ Pro 1700	950–1650 nm and 900–2100 nm	PLS and LDA	Basri et al. (2018)
Horse, chicken, Turkey, beef, and mutton meat	Pork	FT-NIR Büchi NIRFlex N-500	4000 12,500 cm^{-1}/800 2500 nm	PLS-R	Wiedemair et al. (2018)
Ground meat, beef, and chicken		MicroNIR™ Pro 1700	908 to 1676 nm	PLS and SVR	Silva et al. (2020)
Minced beef, lamb, chicken		Cary 630 FTIR spectrometer (Agilent Technologies)	3000–2800 cm^{-1}, 1800–1700, and 1700–1000 cm^{-1}	PCA, PLS, SVM, OOC, and RBF	Dashti et al. (2022)

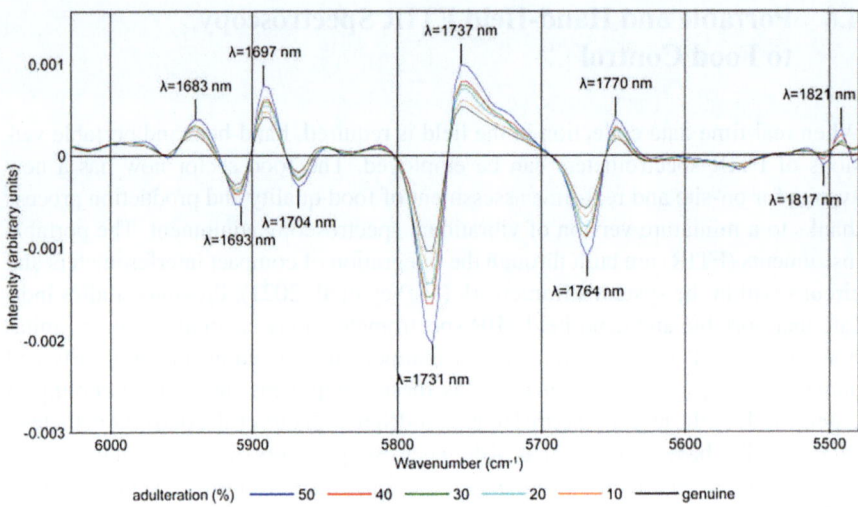

Fig. 4.78 2nd Derivative spectra (6028–5480 cm^{-1}) measured with the laboratory setup as a function of the adulteration of the veal product with pork. Adulteration levels from genuine (no adulteration) up to 50% (in 10% steps). (Schmutzler et al. 2015)

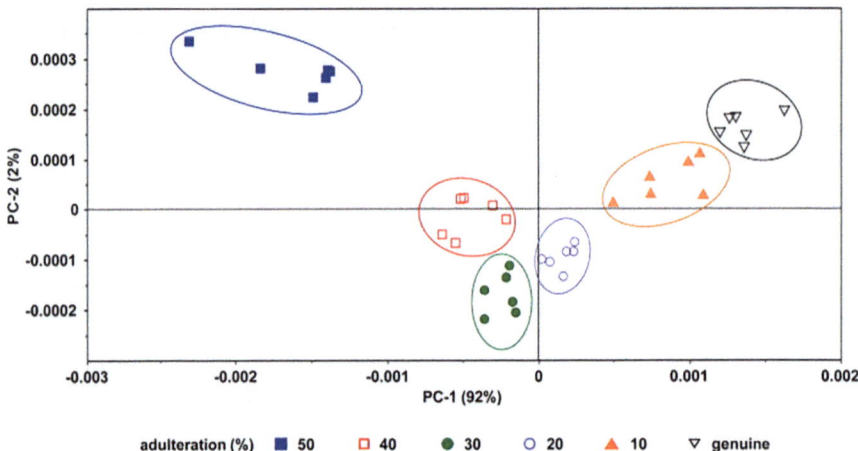

Fig. 4.79 Laboratory setup: PCA scores plot of all adulteration levels and the genuine product for pork detection in veal sausages. Adulteration of the product from genuine up to 50% in 10% steps. (Schmutzler et al. 2015)

PCA scores were processed using SVM classification, enabling precise discrimination of adulterated samples. Laboratory and industrial fiber optic setups detected adulteration down to 10% of contamination, even through quartz and polymer packaging (Figs. 4.79 and 4.80). The handheld spectrometer effectively detected meat adulteration at 10% and fat adulteration at 20% (quartz) and 40% (polymer packaging). The study highlights SVM as a robust chemometric tool for automated food fraud detection, ensuring rapid and reliable meat authentication in diverse settings.

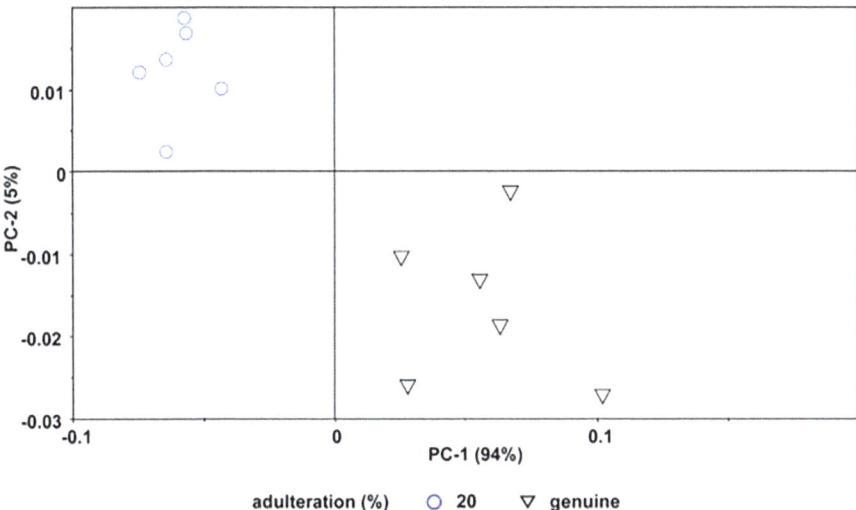

Fig. 4.80 On-site setup: PCA scores plot of samples with 20% adulteration of the fat part and the genuine product. Measurements through quartz cuvettes. (Schmutzler et al. 2015)

In another study, Basri et al. (2018) evaluated the effectiveness of FT-NIR, MicroNIR, and LED-MicroNIR spectroscopy for detecting lard adulteration in palm oil using PLS regression and LDA. Spectral data were acquired from palm oil samples adulterated with lard at various concentrations (0.5%–50%) and processed with different chemometric pretreatments. FT-NIR exhibited the highest spectral resolution (900–2100 nm) and best predictive accuracy ($R^2 = 0.9997$, RMSEP <1), while MicroNIR offered comparable performance with a more compact design. LED-MicroNIR, despite its lower resolution, achieved high classification accuracy (sensitivity = 1.00, specificity = 0.9333), demonstrating the feasibility of LED-based systems for low-cost, real-time adulteration screening. The study highlights NIR spectroscopy as a rapid, reliable, and portable technique for food authenticity and Halal verification. Similarly, Wiedemair et al. (2018) conducted a classification of 63 samples from various meat types, 9 beef, 10 mutton, 10 chicken, 10 turkey, 10 pork, and 14 horse meat, utilizing a portable micro-electromechanical system (MEMS) based spectrometer alongside a bench-top FT-NIR instrument. For the collection of data, portable, and hand-held optical NIR spectra were recorded across the entire range of NIR (12,500 cm^{-1} to 4000 cm^{-1}). The accuracy of MEMS compared to FT-NIR in identifying whole and minced pieces of chicken, pork, turkey, beef, and mutton meat (63 samples) against horse meat was observed to be 75.0–100.0% for MEMS vs. 62.5–100.0% for FT-NIR in whole pieces, while both methods achieved 75.0–100.0% accuracy for minced meat. The quality parameters obtained for the MEMS device included R^2 ranging from 0.06 to 0.62 and SECV values between 17.33 and 32.91. In contrast, the FTNIR system exhibited R^2 ranging from 0.85 to 0.94 and SECV values of 7.52 to 13.83%. The detection limit was established at 10% for the MEMS and 1% for the FT-NIR device. Dashti et al.

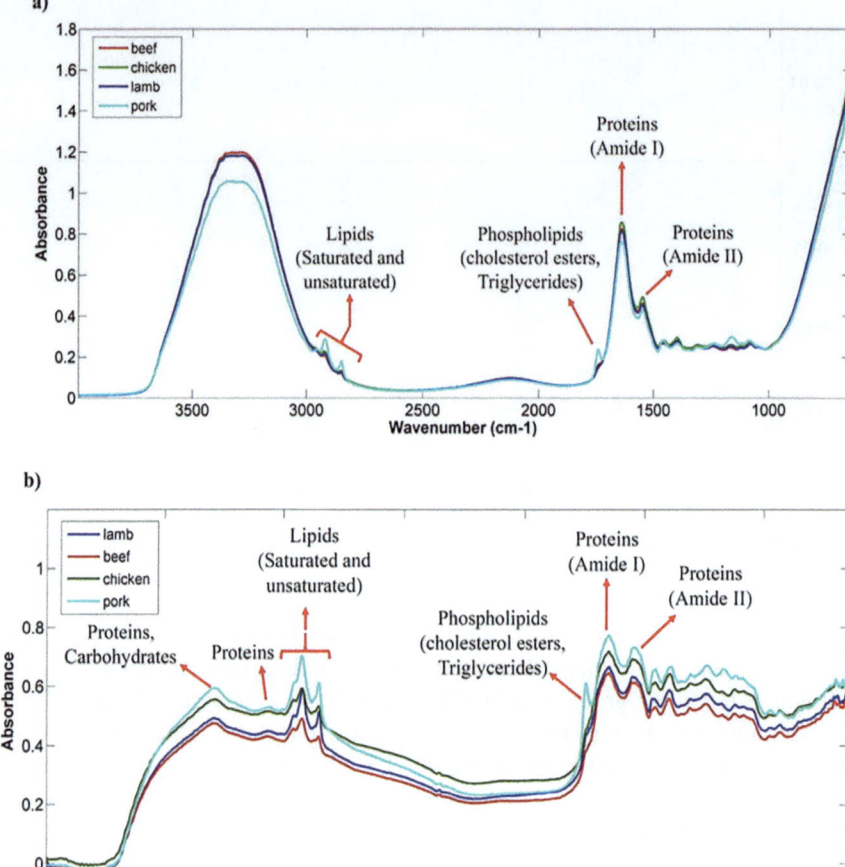

Fig. 4.81 Mean of (**a**) ATR-FTIR spectra and (**b**) DR-FTIR spectra of lamb, beef, chicken, and pork samples. (Dashti et al. 2022)

(2022), in a recent study, evaluated the practicality of using portable FTIR in conjunction with multivariate classification methods for the categorization of various meat samples (chicken, lamb, minced beef, and pork). In this context, both diffuse reflectance (DR) FTIR and ATR-FTIR methods (Fig. 4.81) were assessed within the spectral ranges of 3000–2800 cm^{-1}, 1800–1700, and 1700–1000 cm^{-1}. First, PCA was employed to investigate FT-IR spectra in order to identify differences and similarities between the samples (Fig. 4.82). Additionally, a novel method was used involving one-class classification (OCC) for the certification of Halal meat species. The OCC approach revealed that both DR and ATR methods yielded high false-positive scores in the classification of pork, with the DR method achieving an 89% correct classification rate across all species examined concurrently. The subsequent

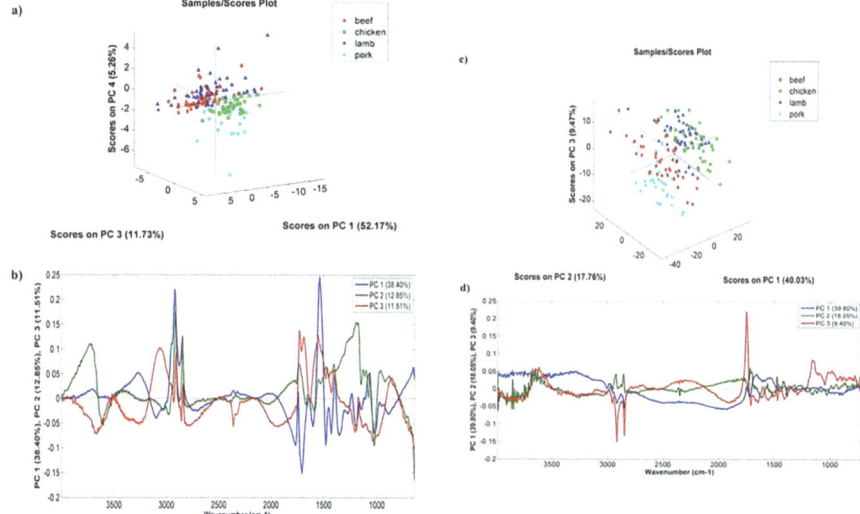

Fig. 4.82 The PCA score and loading plots: (**a**) PCA score plot of ATR-FTIR (3000–2830 and 1760–1000 cm^{-1}) preprocessed with baseline correction (automatic Whittaker filter) + (EPO) external parameter orthogonalization (2PCs), (**b**) PCA loading plot of ATR-FTIR (full spectra) preprocessed with baseline correction (automatic Whittaker filter) + EPO (2PCs), (**c**) PCA score plot of DR-FTIR preprocessed with mean center + EPO (3 PCs), (**d**) PCA loading plot of DR-FTIR (full spectra) preprocessed with mean center + EPO (3 PCs). (Dashti et al. 2022)

phase involved the evaluation of PLS-DA and SVM utilizing the radial basis function (RBF) as the kernel function for meat speciation. The classification performance of SVM was superior in terms of total accuracy, achieving 98% for the ATR-FTIR dataset and 100% for the DR-FTIR datasets, compared to PLS-DA, which recorded 90% and 98%, respectively).

Similarly, Silva et al. (2020) prepared ground meat samples with varying compositions from 0 to 100 wt% in both binary blends (chicken/beef; beef/pork; chicken/pork) and ternary blends (chicken/beef/pork) to identify adulteration in meat through the use of portable NIR spectrometer (Fig. 4.83). In the binary blends, R^2c and R^2p values varied from 0.78 to 0.99, with optimal results obtained for the prediction of chicken content in beef blends (LOD = 3.4 wt% and LOQ 11.2 wt%). The analytical performance of the ternary blends was deemed satisfactory solely for predicting beef content, with a LOD of 4.7 wt% and LOQ of 15.7 wt%. To enhance the analytical performance of the regression models developed, the NIR spectra (binary and ternary blends) were consolidated into a single matrix. SVR and PLS regression models were developed, with the SVR model demonstrating superior performance parameters compared to the PLS model. The portable NIR spectrometer demonstrated effective performance in quantifying beef within ground meat blends, including chicken/beef, pork/beef, and chicken/beef/pork.

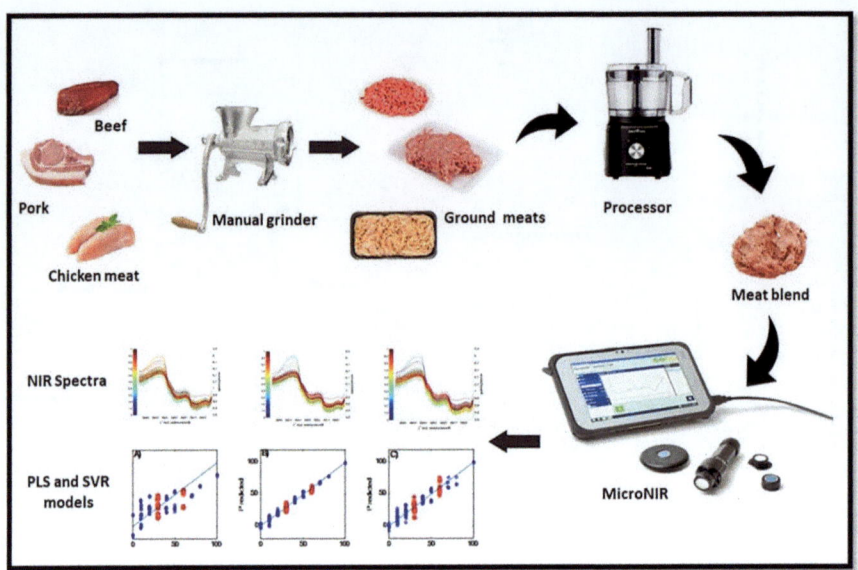

Fig. 4.83 Ground meat (chicken/beef; beef/pork; chicken/pork; chicken/beef/pork) identification using a portable NIR spectrometer. (Silva et al. 2020)

4.9 Current Status, Challenges, and Future Perspectives

Food authentication is crucial for safeguarding the integrity of the global food supply chains amidst evolving industry dynamics. Globalization, the sophistication of food fraud, and consumer expectations are some of the factors that call for strong actions (Wang et al. 2022a, b). The current challenges in food authentication are depicted in Fig. 4.84 (He et al. 2024). This chapter explores the use of IR spectroscopy in verifying the authenticity of Halal food products by analyzing spectral data across various categories. This indicates an encouraging forward direction for research in the implementation of chemometric tools to improve traceability and fraud detection of Halal food products based on robust analysis techniques. The Halal industry has demonstrated massive potential over the past decade, and Halal products are widely recognized in many countries. Several significant research gaps have been addressed. Many models can identify Halal ingredients present in a food product. However, the limited application of chemometric models in research using spectral data is expected to be a contributing factor.

Innovative, reliable, and cost-effective authentication technologies play a significant role. FT-IR spectroscopy offers both qualitative and quantitative insights, making it a powerful tool for analyzing food and biological samples. Its ability to rapidly and accurately assess quality, authenticity, and adulteration in various food products makes it a highly valuable technique. The method stands out for being cost-effective, highly repeatable, and capable of multiple analyses within a single test. With the advancement in instrumentation and chemometric techniques, FTIR spectroscopy

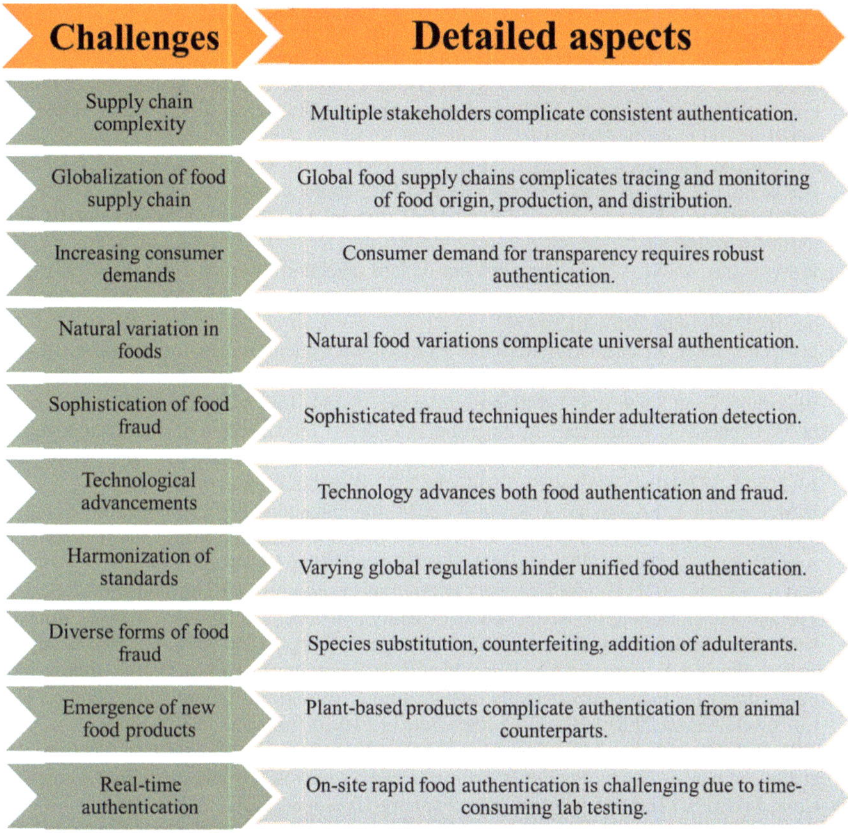

4.9 Current Status, Challenges, and Future Perspectives

Challenges	Detailed aspects
Supply chain complexity	Multiple stakeholders complicate consistent authentication.
Globalization of food supply chain	Global food supply chains complicates tracing and monitoring of food origin, production, and distribution.
Increasing consumer demands	Consumer demand for transparency requires robust authentication.
Natural variation in foods	Natural food variations complicate universal authentication.
Sophistication of food fraud	Sophisticated fraud techniques hinder adulteration detection.
Technological advancements	Technology advances both food authentication and fraud.
Harmonization of standards	Varying global regulations hinder unified food authentication.
Diverse forms of food fraud	Species substitution, counterfeiting, addition of adulterants.
Emergence of new food products	Plant-based products complicate authentication from animal counterparts.
Real-time authentication	On-site rapid food authentication is challenging due to time-consuming lab testing.

Fig. 4.84 Current challenges proposed for addressing in the future study

has seen significant growth in industrial applications, particularly in quality control and process monitoring (Kumar et al. 2023; Wu and Kamruzzaman 2024). Despite its many advantages, FT-IR spectroscopy faces intrinsic limitations that hinder its widespread industrial application. As an indirect method reliant on accurate wet chemistry analysis, FT-IR requires complex multivariate calibration. Effective multivariate data mining techniques are essential for enhancing predictive accuracy, with variable selection playing a key role in improving model interpretability, boosting prediction performance, and accelerating processing. Future advancements in FT-IR technology, including the use of selected wavelengths and enhanced hardware, alongside the development of robust algorithms, are expected to support real-time inspection of quality, authenticity, and adulteration detection (Ray et al. 2022; Guan et al. 2023). A multidisciplinary approach integrating technical innovation, regulatory collaboration, and stakeholder engagement is crucial for addressing the complexities of the modern food industry, ensuring consumer demands are met while upholding food safety, authenticity, and the integrity of the global food supply chain.

References

Abbas O, Fernández Pierna JA, Codony R, von Holst C, Baeten V (2009) Assessment of the discrimination of animal fat by FT-Raman spectroscopy. J Mol Struct 924:294–300. https://doi.org/10.1016/j.molstruc.2009.01.027

Abu-Ghoush M, Fasfous I, Al-Degs Y, Al-Holy M, Issa AA, Al-Reyahi AY, Alshathri AA (2017) Application of mid-infrared spectroscopy and PLS-kernel calibration for quick detection of pork in higher value meat mixes. J Food Meas Charact 11:337–346. https://doi.org/10.1007/s11694-016-9402-4

Ahda M, Safitri ANNA (2016) Development of lard detection in crude palm oil (CPO) using ftir combined with chemometrics analysis. Int J Pharm Pharm Sci 8:307–309. https://doi.org/10.22159/ijpps.2016v8i12.14743

Ahda M, Guntarti A, Kusbandari A, Melianto Y (2020) Authenticity analysis of beef meatball adulteration with wild boar using FTIR spectroscopy combined with chemometrics. J Microbiol Biotechnol Food Sci 9(5):937–940. https://doi.org/10.15414/jmbfs.2020.9.5.937-940

Ahda M, Guntarti A, Kusbandari A, Nugroho HA (2023) Identification of adulterated sausage products by pork using FTIR and GC-MS combined with chemometrics. J Chem Health Risks 13(2):325–332. https://doi.org/10.22034/jchr.2022.1946997.1465

Ahda M, Guntarti A, Kusbandari A, Safitri A (2024) Identification of lard adulteration of cooking oil products using Fourier transform infrared spectroscopy combined with chemometrics. Food Anal Methods 17(3):366–372. https://doi.org/10.1007/s12161-024-02576-y

Ahmad A, Ayub H (2022) Fourier transform infrared spectroscopy (FTIR) technique for food analysis and authentication. In: Nondestructive quality assessment techniques for fresh fruits and vegetables. Springer, Singapore, pp 103–142. https://doi.org/10.1007/978-981-19-5422-1_6

Ahmad S, Wong KY, Rashid AFA, Khan M (2024) Environmental impacts and improvement implications for industrial meatballs manufacturing: scenario in a developing country. Int J Life Cycle Assess 29(8):1510–1522. https://doi.org/10.1007/s11367-023-02146-0

Ali ME, Nina Naquiah AN, Mustafa S, Hamid SBA (2015) Differentiation of frog fats from vegetable and marine oils by Fourier transform infrared spectroscopy and chemometric analysis. Croat J Food Sci Technol 7(1):1–8. https://doi.org/10.17508/CJFST.2015.7.1.03

Al-Jowder OA, Kemsley EK, Wilson RH (1997) Mid-infrared spectroscopy and authenticity problems in selected meats: a feasibility study. Food Chem 59(2):195–201. https://doi.org/10.1016/S0308-8146(96)00289-0

Al-Jowder O, Defernez M, Kemsley EK, Wilson RH (1999) Mid-infrared spectroscopy and chemometrics for the authentication of meat products. J Agric Food Chem 47:3210–3218. https://doi.org/10.1021/jf981196d

Alkhalf MI, Mirghani ME, Nazrim Marikkar JM, Hammed AM, Kabbashi NA (2017) The use of analytical techniques for qualitative differentiation of lipids extracted from cheese samples and lard. J Food Agric Environ 15:20–25

Anggarkasih MG, Resma PS (2022) The importance of Halal certification for the processed food by SMEs to increase export opportunities. In: E3S web of conferences, vol 348. EDP Sci, p 00039. https://doi.org/10.1051/e3sconf/202234800039

Arifah MF, Irnawati R, Nisa K, Windarsih A, Rohman A (2022) The application of FTIR spectroscopy and chemometrics for the authentication analysis of horse milk. Int J Food Sci 2022(1):7643959. https://doi.org/10.1155/2022/7643959

Arnalds T, McElhinney J, Fearn T, Downey G (2004) A hierarchical discriminant analysis for species identification in raw meat by visible and near infrared spectroscopy. J Near Infrared Spectrosc 12(3):183–188. https://doi.org/10.1255/jnirs.425

Augustyńska-Prejsnar A, Sokołowicz Z, Ormian M, Tobiasz-Salach R (2022) Nutritional and health-promoting value of poultry meatballs with the addition of plant components. Foods 11(21):3417. https://doi.org/10.3390/foods11213417

References

Aziz AA, Sani MSA, Zakaria Z, Bakar NKA (2023) Discrimination and authentication of lard blending with palm oil in cosmetic soap formulations. Int J Cosmet Sci 45(4):444–457. https://doi.org/10.1111/ics.12854

Bahri SS, Che Man YB (2016) Rapid detection of lard in chocolate and chocolate-based food products using Fourier transform infrared spectroscopy

Basri KN, Laili AR, Tuhaime NA, Hussain MN, Bakar J, Sharif Z, Khir MFA, Zoolfakar AS (2018) FT-NIR, MicroNIR and LED-MicroNIR for detection of adulteration in palm oil via PLS and LDA. Anal Methods 10(34):4143–4151. https://doi.org/10.1039/c8ay01239c

Beć KB, Huck CW (2019) Breakthrough potential in near-infrared spectroscopy: spectra simulation. A review of recent developments. Front Chem 7:48. https://doi.org/10.3389/fchem.2019.00048

Biancolillo A, Marini F, Ruckebusch C, Vitale R (2020) Chemometric strategies for spectroscopy-based food authentication. Appl Sci 10(18):6544. https://doi.org/10.3390/app10186544

Candoğan K, Altuntas EG, İğci N (2021) Authentication and quality assessment of meat products by Fourier-transform infrared (FTIR) spectroscopy. Food Eng Rev 13(1):66–91. https://doi.org/10.1007/s12393-020-09251-y

Cebi N, Durak MZ, Toker OS, Sagdic O, Arici M (2016) An evaluation of Fourier transforms infrared spectroscopy method for the classification and discrimination of bovine, porcine and fish gelatins. Food Chem 190:1109–1115. https://doi.org/10.1016/j.foodchem.2015.06.065

Cebi N, Dogan CE, Mese AE, Ozdemir D, Arici M, Sagdic O (2019) A rapid ATR-FTIR spectroscopic method for classification of gelatin gummy candies in relation to the gelatin source. Food Chem 277:373–381. https://doi.org/10.1016/j.foodchem.2018.10.125

Cebi N, Bekiroglu H, Erarslan A, Rodriguez-Saona L (2023) Rapid sensing: hand-held and portable FTIR applications for on-site food quality control from farm to fork. Molecules 28(9):3727. https://doi.org/10.3390/molecules28093727

Che Man YB, Mirghani MES (2001) Detection of lard mixed with body fats of chicken, lamb, and cow by Fourier transform infrared spectroscopy. J Am Oil Chem Soc 78(7):753–761. https://doi.org/10.1007/s11746-001-0338-4

Che Man YB, Rohman A (2011) Detection of lard in vegetable oils. Lipid Technol 23(8):180–182. https://doi.org/10.1002/lite.201100128

Che Man YB, Syahariza Z a, Mirghani MES, Jinap S, Bakar J (2005) Analysis of potential lard adulteration in chocolate and chocolate products using Fourier transform infrared spectroscopy. Food Chem 90(4):815–819. https://doi.org/10.1016/j.foodchem.2004.05.029

Che Man YB, Syahariza Z a, Rohman A (2011a) Discriminant analysis of selected edible fats and oils and those in biscuit formulation using FTIR spectroscopy. Food Anal Methods 4:404–409. https://doi.org/10.1007/s12161-010-9184-y

Che Man YB, Rohman A, Mansor TST (2011b) Differentiation of lard from other edible fats and oils by means of Fourier transform infrared spectroscopy and chemometrics. J Am Oil Chem Soc 88(2):187–192. https://doi.org/10.1007/s11746-010-1659-x

Che Man YB, Marina AM, Rohman A, Al-Kahtani HA, Norazura O (2014) A Fourier transform infrared spectroscopy method for analysis of palm oil adulterated with lard in pre-fried French fries. Int J Food Prop 17(2):354–362. https://doi.org/10.1080/10942912.2011.631254

Cozzolino D, Murray I (2004) Identification of animal meat muscles by visible and near infrared reflectance spectroscopy. LWT 37(4):447–452. https://doi.org/10.1016/j.lwt.2003.10.013

da Costa Fulgêncio AC, Resende GAP, Teixeira MCF, Botelho BG, Sena MM (2023) Combining portable NIR spectroscopy and multivariate calibration for the determination of ethanol in fermented alcoholic beverages by a multi-product model. Talanta Open 7:100180

Dama RW, Nainggolan B, Rahmat R (2025) Alcoholic beverage transactions in the perspective of Maqasid Shariah: a traditional approach in ternate culture. In: International conference on research issues and community service, vol 1(1), pp 312–320. https://doi.org/10.31332/i-cores.v1i1.11246

Danezis GP, Tsagkaris AS, Camin F, Brusic V, Georgiou CA (2016) Food authentication: techniques, trends & emerging approaches. TrAC Trends Anal Chem 85:123–132. https://doi.org/10.1016/j.trac.2016.02.026

Dashti A, Weesepoel Y, Müller-Maatsch J, Parastar H, Kobarfard F, Daraei B, Yazdanpanah H (2022) Assessment of meat authenticity using portable Fourier transform infrared spectroscopy combined with multivariate classification techniques. Microchem J 181:107735. https://doi.org/10.1016/j.microc.2022.107735

De Cicco M, Siano F, Iacomino G, Iannaccone N, Di Stasio L, Mamone G, Volpe MG, Ferranti P, Addeo F, Picariello G (2019) Multianalytical detection of pig-derived ingredients in bread. Food Anal Methods 12:780–790. https://doi.org/10.1007/s12161-018-01410-6

Debebe A, Anberbir A, Redi-Abshiro M, Chandravanshi BS, Asfaw A, Asfaw N, Retta N (2017a) Alcohol determination in distilled alcoholic beverages by liquid phase Fourier transform mid-infrared and near-infrared spectrophotometries. Food Anal Methods 10:172–179. https://doi.org/10.1007/s12161-016-0566-7

Debebe A, Temesgen S, Redi-Abshiro M, Chandravansh BS (2017b) Partial least squares− near infrared spectrometric determination of ethanol in distilled alcoholic beverages. Bull Chem Soc Ethiop 31(2):201–209. https://doi.org/10.4314/bcse.v31i2.2

Debebe ADA, Redi-Abshiro M, Chandravanshi BS (2017c) Non-destructive determination of ethanol levels in fermented alcoholic beverages using Fourier transform mid-infrared spectroscopy. Chem Cent J 11:27. https://doi.org/10.1186/s13065-017-0257-5

Di Donato F, Biancolillo A, Ferretti A, D'Archivio AA, Marini F (2023) Near infrared spectroscopy coupled to chemometrics for the authentication of donkey milk. J Food Compos Anal 115:105017. https://doi.org/10.1016/j.jfca.2022.105017

Ding HB, Xu RJ (2000) Near-infrared spectroscopic technique for detection of beef hamburger adulteration. J Agric Food Chem 48(6):2193–2198. https://doi.org/10.1021/jf9907182

Djekic I, Smigic N (2023) Consumer perception of food fraud in Serbia and Montenegro. Foods 13(1):53. https://doi.org/10.3390/foods13010053

DSM (2008) Malaysian standard MS 2200: part 1:2008 Islamic consumer goods—part 1: cosmetic and personal care – general guidelines. Department of Standards Malaysia, Shah Alam, pp 1–6. Printing Department, SIRIM Berhad

Du C, Zhou J (2009) Evaluation of soil fertility using infrared spectroscopy: a review. Environ Chem Lett 7:97–113. https://doi.org/10.1007/s10311-008-0166-x

Dutta A (2017) Fourier transform infrared spectroscopy. In: Spectroscopic methods for nanomaterials characterization, pp 73–93. https://doi.org/10.1016/B978-0-323-46140-5.00004-2

Eckl MR, Biesbroek S, Veer P v't, Geleijnse JM (2021) Replacement of meat with non-meat protein sources: a review of the drivers and inhibitors in developed countries. Nutrients 13(10):3602. https://doi.org/10.3390/nu13103602

El Sheikha AF, Mokhtar NFK, Amie C, Lamasudin DU, Isa NM, Mustafa S (2017) Authentication technologies using DNA-based approaches for meats and halal meats determination. Food Biotechnol 31(4):281–315. https://doi.org/10.1080/08905436.2017.1369886

Ellis DI, Broadhurst D, Clarke SJ, Goodacre R (2005) Rapid identification of closely related muscle foods by vibrational spectroscopy and machine learning. Analyst 130:1648–1654. https://doi.org/10.1039/B511484E

Erwanto Y, Muttaqien AT, Sugiyono S, Rohman A (2016) Use of Fourier transform infrared (FTIR) spectroscopy and chemometrics for analysis of lard adulteration in "rambak" crackers. Int J Food Prop 19(12):2718–2725. https://doi.org/10.1080/10942912.2016.1143839

EU (1976) EU cosmetic directive 76/768/EEC. The European Union

Fajriati I, Rosadi Y, Rosadi NN, Khamidinal. (2021) Detection of animal fat mixtures in meatballs using fourier transform infrared spectroscopy (FTIR spectroscopy). Indones J Halal Res 3(1):8–12. https://doi.org/10.15575/ijhar.v3i1.11166

Fan Y, Cheng F, Xie L (2010) Quantitative analysis and detection of adulteration in pork using near-infrared spectroscopy. In: Sensing for agriculture and food quality and safety II, vol 7676. SPIE, pp 230–237. https://doi.org/10.1117/12.852521

FAO, Food Systems Assessment Programme (2014) Health division: meat & meat products. Food and Agricultural Organization of the United Nations, Rome. http://www.fao.org/ag/againfo/themes/en/meat/background.html

Fatmawati S, Hariyanti H, Permanasari ED, Chandra A, Firmanto JD, Sari RL, Pamungkas ST (2024) Determination of gelatin from marshmallows using a combination of Fourier transform infrared (ATR-FTIR) spectroscopic and chemometrics for halal authentication. J Halal Sci Res 5(2):149–155. https://doi.org/10.12928/jhsr.v5i2.10537

Fengou L-C, Lianou A, Tsakanikas P, Mohareb F, Nychas G-JE (2021) Detection of meat adulteration using spectroscopy-based sensors. Foods 10(4):861. https://doi.org/10.3390/foods10040861

Coldea TE, Socaciu C, Fetea F, Ranga F, Pop RM, Florea M (2013) Rapid quantitative analysis of ethanol and prediction of methanol content in traditional fruit brandies from Romania, using FTIR spectroscopy and chemometrics. Notulae Botanicae Horti Agrobotanici Cluj-Napoca 41(1):143–149. https://doi.org/10.15835/nbha4119000

Flores M, Piornos JA (2021) Fermented meat sausages and the challenge of their plant-based alternatives: a comparative review on aroma-related aspects. Meat Sci 182:108636. https://doi.org/10.1016/j.meatsci.2021.108636

Gallignani M, Ayala C, Brunetto Mdel R, Burguera JL, Burguera M (2005) A simple strategy for determining ethanol in all types of alcoholic beverages based on its on-line liquid–liquid extraction with chloroform, using a flow injection system and Fourier transform infrared spectrometric detection in the mid-IR. Talanta 68(2):470–479. https://doi.org/10.1016/j.talanta.2005.09.031

Gangidi RR, Proctor A, Pohlman FW (2003) Rapid determination of spinal cord content in ground beef by attenuated total reflectance Fourier transform infrared spectroscopy, J Food Sci 68:124–127. https://doi.org/10.1111/j.1365-2621.2003.tb14126.x

Gao B, Xu S, Han L, Liu X (2020) FT-IR-based quantitative analysis strategy for target adulterant in fish oil multiply adulterated with terrestrial animal lipid. Food Chem 343:128420. https://doi.org/10.1016/j.foodchem.2020.128420

Guan H, Shiqin D, Han B, Zhang Q, Wang D (2023) A rapid and sensitive smartphone colorimetric sensor for detection of ascorbic acid in food using the nanozyme paper-based microfluidic chip. LWT 184:115043. https://doi.org/10.1016/j.lwt.2023.115043

Guillén MD, Cabo N (1997) Characterization of edible oils and lard by Fourier transform infrared spectroscopy. Relationships between composition and frequency of concrete bands in the fingerprint region. J Am Oil Chem Soc 74(10):1281–1286. https://doi.org/10.1007/s11746-997-0058-4

Guillén MD, Cabo N (1999) Usefulness of the frequency data of the Fourier transform infrared spectra to evaluate the degree of oxidation of edible oils. J Agric Food Chem 47(2):709–719. https://doi.org/10.1021/jf9808123

Guntarti A, Prativi SR (2017) Application method of Fourier transform infrared (FTIR) combined with chemometrics for analysis of rat meat (Rattus Diardi) in meatballs beef. Pharmaciana 7(2):133. https://doi.org/10.12928/pharmaciana.v7i2.4247

Guntarti A, Purbowati ZA (2019) Analysis of dog fat in beef sausage using FTIR (Fourier transform infrared) combined with chemometrics. Pharmaciana 9(9):21–28. https://doi.org/10.12928/pharmaciana.v9i1.10467

Guntarti A, Martono S, Yuswanto A, Rohman A (2015) FTIR spectroscopy in combination with chemometrics for analysis of wild boar meat in meatball formulation. Asian J Biochem 10(4):165–172. https://doi.org/10.3923/ajb.2015.165.172

Guntarti A, Ahda M, Kusbandari A, Sauri AS (2018) Fourier-transform infrared spectroscopy combined with chemometrics for detection of pork in beef meatball formulation. Int J Green Pharm 12(3):153. https://doi.org/10.22377/ijgp.v12i03.1945

Guntarti A, Ahda M, Kusbandari A, Prihandoko SW (2019a) Analysis of lard in sausage using Fourier transform infrared spectrophotometer combined with chemometrics. Int J Pharm Bio Sci 11(Suppl 4):S594–S600. https://doi.org/10.4103/jpbs.JPBS_209_19

Guntarti A, Ahda M, Sunengsih N (2019b) Identification of lard on grilled beef sausage product and steamed beef sausage product using Fourier transform infrared (FTIR) spectroscopy with chemometric combination. Potravinarstvo Slovak J Food Sci 13(1):767–772. https://doi.org/10.5219/1162

Guntarti A, Ahda M, Kusbandari A, Natalie F (2020) Analysis of pork adulteration in the corned products using FTIR associated with chemometrics analysis. Slovak J Food Sci/Potravinarstvo 14(1). https://doi.org/10.5219/1412

Habiba U, Hossain MM, Habib M, Hashem MA, Ali MS (2021) Effect of adding different types of flour on the quality of low fat beef sausage. Bangladesh J Anim Sci 50(1):1–11

Haider A, Iqbal SZ, Bhatti IA, Alim MB, Waseem M, Iqbal M, Khaneghah AM (2024) Food authentication, current issues, analytical techniques, and future challenges: a comprehensive review. Compr Rev Food Sci Food Saf 23(3):e13360. https://doi.org/10.1111/1541-4337.13360

Hashim DM, Che Man YB, Norakasha R, Shuhaimi MB, Salmah Y, Syahariza ZA (2010) Potential use of Fourier transform infrared spectroscopy for differentiation of bovine and porcine gelatins. Food Chem 118(3):856–860. https://doi.org/10.1016/j.foodchem.2009.05.049

Hassan N, Ahmad T, Zain NM (2018) Chemical and chemometric methods for halal authentication of gelatin: an overview. J Food Sci 83(12):2903–2911. https://doi.org/10.1111/1750-3841.14370

Hassan N, Ahmad T, Zain NM, Awang SR (2021) Identification of bovine, porcine and fish gelatin signatures using chemometrics fuzzy graph method. Sci Rep 11(1):9793. https://doi.org/10.1038/s41598-021-89358-2

He HJ, da Silva Ferreira MV, Wu Q, Karami H, Kamruzzaman M (2024) Portable and miniature sensors in supply chain for food authentication: a review. Crit Rev Food Sci Nutr 1–21. https://doi.org/10.1080/10408398.2024.2380837

Henchion MM, McCarthy M, Resconi VC (2017) Beef quality attributes: a systematic review of consumer perspectives. Meat Sci 128:1–7. https://doi.org/10.1016/j.meatsci.2017.01.006

Herrero AM, Cambero MI, Ordóñez JA, de la Hoz L, Carmona P (2009) Plasma powder as cold-set binding agent for meat system: rheological and Raman spectroscopy study. Food Chem 113(2):493–499. https://doi.org/10.1016/j.foodchem.2008.07.084

Hossaina MAM, Uddin SMK, Sultana S, Wahab YA, Sagadevan S, Johan MR, Eaqub Ali M (2021) Authentication of Halal and Kosher meat and meat products: analytical approaches, current progresses and future prospects. Crit Rev Food Sci Nutr 62(2):285–310. https://doi.org/10.1080/10408398.2020.1814691

Hu Y, Zou L, Huang X, Xiaonan L (2017) Detection and quantification of offal content in ground beef meat using vibrational spectroscopic-based chemometric analysis. Sci Rep 7(1):15162. https://doi.org/10.1038/s41598-017-15389-3

Huda N, Shen YH, Huey YL, R Ahmad., and A, Mardiah. (2010) Evaluation of physico-chemical properties of Malaysian commercial beef meatballs. Am J Food Technol 5(1):13–21

Irnawati I, Windarsih A, Indrianingsih WA, Apriyana W, Ratnawati YA, Nadia LOMH, Rohman A (2023) Rapid detection of tuna fish oil adulteration using FTIR-ATR spectroscopy and chemometrics for halal authentication. J Appl Pharm Sci 13(4):231–239. https://doi.org/10.7324/JAPS.2023.120270

Jaswir I, Mirghani MES, Hassan TH, Said MZM (2003) Determination of lard in mixture of body fats of mutton and cow by Fourier transform infrared spectroscopy. J Oleo Sci 52(12):633–638. https://doi.org/10.5650/jos.52.633

Karabagias IK (2020) Advances of spectrometric techniques in food analysis and food authentication implemented with chemometrics. Foods 9(11):1550. https://doi.org/10.3390/foods9111550

Karim NA, Muhamad II (2018) Detection methods and advancement in analysis of food and beverages: a short review on adulteration and Halal authentication. In: Proceedings of the 3rd international halal conference (INHAC 2016). Springer, Singapore, pp 397–414

Kazazić S, Gajdoš-Kljusurić J, Radeljević B, Plavljanić D, Špoljarić J, Ljubić T, Bilić B, Mikulec N (2021) Comparison of GC and NIR spectra as a rapid tool for food fraud detection: case of butter adulteration with different fat types. J Food Process Preserv 45(9):e15732. https://doi.org/10.1111/jfpp.15732

Kendall H, Clark B, Rhymer C, Kuznesof S, Hajslova J, Tomaniova M, Brereton P, Frewer L (2019) A systematic review of consumer perceptions of food fraud and authenticity: a European perspective. Trends Food Sci Technol 94:79–90. https://doi.org/10.1016/j.tifs.2019.10.005

Khattak H (2009) Halal certified cosmetics and personal care products—where purity comes first. Halal Dig 1:1–3. Retrieved on February, 2011 from http://www.infanca.org/newsletter/2009.01.htm

Kowalska A (2018) The study of the intersection between food fraud/adulteration and authenticity. Acta Univ Agric Silvic Mendelianae Brun 66(5):1275–1286. https://doi.org/10.11118/actaun201866051275

Kuang L, Tian X, Ying S, Chen Chen L, Zhao XM, Han L, Chen C, Zhang J (2025) Rapid identification of horse oil adulteration based on deep learning infrared spectroscopy detection method. Spectrochim Acta A Mol Biomol Spectrosc 330:125604. https://doi.org/10.1016/j.saa.2024.125604

Kumar A, Tripathi S, Hans N, Pattnaik HSN, Naik SN (2018) Ghee: its properties, importance and health benefits. Lipid Universe 6:6–14

Kumar A, Castro M, Feller J-F (2023) Review on sensor array-based analytical technologies for quality control of food and beverages. Sensors 23(8):4017. https://doi.org/10.3390/s23084017

Kunbhar S, Talpur FN, Mahesar SA, Afridi HI, Fareed G, Razzaque N, Nisa M-u (2025) Application of ATR-FTIR and chemometrics for rapid lard adulteration assessment in confectionery. Vib Spectrosc 136:103762. https://doi.org/10.1016/j.vibspec.2024.103762

Kurniadi M, Frediansyah A (2017) Halal perspective of microbial bioprocess based-food products. Reaktor 16(3):147–160. https://doi.org/10.14710/reaktor.16.3.147-160

Kurniawati E, Rohman A, Triyana K (2014) Analysis of lard in meatball broth using Fourier transform infrared spectroscopy and chemometrics. Meat Sci 96(1):94–98. https://doi.org/10.1016/j.meatsci.2013.07.003

Kuswandi B, Putri FK, Gani AA, Ahmad M (2015a) Application of class-modelling techniques to infrared spectra for analysis of pork adulteration in beef jerkys. J Food Sci Technol 52:7655–7668. https://doi.org/10.1007/s13197-015-1882-4

Kuswandi B, Cendekiawan KA, Kristiningrum N, Ahmad M (2015b) Pork adulteration in commercial meatballs determined by chemometric analysis of NIR spectra. J Food Meas Charact 9:313–323. https://doi.org/10.1007/s11694-015-9238-3

Kwasek K, Thorne-Lyman AL, Phillips M (2020) Can human nutrition be improved through better fish feeding practices? A review paper. Crit Rev Food Sci Nutr 60(22):3822–3835. https://doi.org/10.1080/10408398.2019.1708698

Lachenmeier DW (2007) Rapid quality control of spirit drinks and beer using multivariate data analysis of Fourier transform infrared spectra. Food Chem 101(2):825–832. https://doi.org/10.1016/j.foodchem.2005.12.032

Lachenmeier DW, Richling E, López MG, Frank W, Schreier P (2005) Multivariate analysis of FTIR and ion chromatographic data for the quality control of tequila. J Agric Food Chem 53(6):2151–2157. https://doi.org/10.1021/jf048637f

Lamyaa MA (2013) Discrimination of pork content in mixtures with raw minced camel and buffalo meat using FTIR spectroscopic technique. Int Food Res J 20(3):1389

Latif IA, Mohamed Z, Sharifuddin J, Abdullah AM, Ismail MM (2014) A comparative analysis of global halal certification requirements. J Food Prod Mark 20(sup1):85–101. https://doi.org/10.1080/10454446.2014.921869

Ledesma E, Laca A, Rendueles M, Díaz M (2016) Texture, colour and optical characteristics of a meat product depending on smoking time and casing type. LWT 65:164–172. https://doi.org/10.1016/j.lwt.2015.07.077

Leng T, Li F, Xiong L, Xiong Q, Zhu M, Chen Y (2020) Quantitative detection of binary and ternary adulteration of minced beef meat with pork and duck meat by NIR combined with chemometrics. Food Control 113:107203

Lestari D, Rohman A, Syofyan S, Yuliana ND, Bt. Abu Bakar NK, Hamidi D (2022) Analysis of beef meatballs with rat meat adulteration using Fourier transform infrared (FTIR) spectroscopy in combination with chemometrics. Int J Food Prop 25(1):1446–1457. https://doi.org/10.108 0/10942912.2022.2083637

Lestari D, Syamsul ES, Wirnawati, Syafri S, Syofyan S, Rohman A, Yuliana ND, Bakar NKBA, Hamidi D (2024) Rapid detection of rat meat adulteration in beef sausages using FTIR-ATR spectroscopy and chemometrics for halal authentication. Int J Appl Pharm 16(1):82–88. https://doi.org/10.22159/ijap.2024.v16s1.21

Li B, Kelly AL, Mcsweeney PLH (2023., ch. 4,) In: Tunick MH (ed) Handbook of cheese chemistry. The Royal Society of Chemistry, pp 62–86

Lingzhi X, Fei G, Zengling Y, Lujia H, Xian L (2016) Discriminant analysis of terrestrial animal fat and oil adulteration in fish oil by infrared spectroscopy. Int J Agric Biol Eng 9(3):179–185. https://doi.org/10.3965/j.ijabe.20160903.2239

Lukitaningsih E, Sa'adah M, Purwanto P, Rohman A (2012) Quantitative analysis of lard in cosmetic lotion formulation using FTIR spectroscopy and partial least square calibration. J Am Oil Chem Soc 89(8):1537–1543. https://doi.org/10.1007/s11746-012-2052-8

Mabood F, Boqué R, Alkindi AY, Al-Harrasi A, Al Amri IS, Boukra S, Jabeen F, Hussain J, Abbas G, Naureen Z, QMI H, Shah HH, Khan A, Khalaf SK, Kadim I (2020) Fast detection and quantification of pork meat in other meats by reflectance FT-NIR spectroscopy and multivariate analysis. Meat Sci 163:108084. https://doi.org/10.1016/j.meatsci.2020.108084

Mansor TST, Che Man YB, Rohman A (2011) Application of fast gas chromatography and Fourier transform infrared spectroscopy for analysis of lard adulteration in virgin coconut oil. Food Anal Methods 4(3):365–372. https://doi.org/10.1007/s12161-010-9176-y

McVey C, Christopher TE, Cannavan A, Simon DK, Petchkongkaew A, Haughey SA (2021) Portable spectroscopy for high throughput food authenticity screening: advancements in technology and integration into digital traceability systems. Trends Food Sci Technol 118:777–790. https://doi.org/10.1016/j.tifs.2021.11.003

Mikhlin YV, Avramov KV (2015) Review of applications of nonlinear normal modes for vibrating mechanical systems. Appl Mech Rev 65(2):020801

Mohan MS, O'Callaghan TF, Kelly P, Hogan SA (2021) Milk fat: opportunities, challenges and innovation. Crit Rev Food Sci Nutr 61(14):2411–2443. https://doi.org/10.1080/10408398.2020.1778631

Montowska M, Pospiech E (2010) Authenticity determination of meat and meat products on the protein and DNA basis. Food Rev Intl 27(1):84–100. https://doi.org/10.1080/87559129.2010.518297

Munir F, Musharraf SG, Sherazi STH, Malik MI, Bhanger MI (2019) Detection of lard contamination in five different edible oils by FT-IR spectroscopy using a partial least squares calibration model. Turk J Chem 43(4):1098–1108. https://doi.org/10.3906/kim-1902-17

Mursyidi A (2013) The role of chemical analysis in the halal authentication of food and pharmaceutical products. J Food Pharm Sci 1:1–4. https://doi.org/10.14499/jfps

Murugaiah C, Noor ZM, Mastakim M, Bilung LM, Radu JSS (2009) Meat species identification and halal authentication analysis using mitochondrial DNA. Meat Sci 83(1):57–61. https://doi.org/10.1016/j.meatsci.2009.03.015

Muttaqien AT, Erwanto Y, Rohman A (2016) Determination of buffalo and pig "rambak" crackers using FTIR spectroscopy and chemometrics. Asian J Anim Sci 10(1):49–58

Nakyinsige K, Man YBC, Sazili AQ (2012) Review: halal authenticity issues in meat and meat products. Meat Sci 91:207–214. https://doi.org/10.1016/j.meatsci.2012.02.015

Ng PC, Ruslan NASA, Chin LX, Ahmad M, Hanifah SA, Abdullah Z, Khor SM (2022) Recent advances in halal food authentication: challenges and strategies. J Food Sci 87(1):8–35. https://doi.org/10.1111/1750-3841.15998

Thuy Chu•Tan Nguyen, Hyunsang Yoo, Jihoon Wang. (2024). A review of vibration analysis and its applications. Heliyon, 10(5) e26282. https://doi.org/10.1016/j.heliyon.2024.e26282

Nina Naquiah AN, Marikkar JMN, Mirghani MES, Nurrulhidayah AF, Yanty NAM (2017) Differentiation of fractionated components of lard from other animal fats using different analytical techniques. Sains Malays 46(2):209–216. https://doi.org/10.17576/jsm-2017-4602-04

References

Nunes KM, Marcus VO, Andrade AMP, Filho S, Lasmar MC, Sena MM (2016) Detection and characterisation of frauds in bovine meat in natura by non-meat ingredient additions using data fusion of chemical parameters and ATR-FTIR spectroscopy. Food Chem 205:14–22. https://doi.org/10.1016/j.foodchem.2016.02.158

Nurani LH, Kusbandari A, Guntarti A, Ahda M, Warsi FN, Mubarok A, Rohman A (2022) Determination of pork adulteration in roasted beef meatballs using Fourier transform infrared spectroscopy in combination with chemometrics. Sains Malays 51(8):2573–2582. https://doi.org/10.17576/jsm-2022-5108-17

Nurrulhidayah AF, Che Man YB, Amin I, Arieff Salleh R, Yusof FM, Shuhaimi M, Khatib A (2015) FTIR-ATR spectroscopy-based metabolite fingerprinting as a direct determination of butter adulterated with lard. Int J Food Prop 18(2):372–379. https://doi.org/10.1080/10942912.2012.692224

Owusu-Ansah P, Besiwah EK, Bonah E, Amagloh FK (2022) Non-meat ingredients in meat products: a scoping review. Appl Food Res 2(1):100044. https://doi.org/10.1016/j.afres.2022.100044

Panchal B, Bhandari B (2020) Butter and dairy fat spreads. In: Dairy fat products and functionality: fundamental science and technology, pp 509–532. https://doi.org/10.1007/978-3-030-41661-4_21

Pavia DL, Lampman GM, Kriz GS, Vyvyan JA (2008) Introduction to spectroscopy, 4th edn. Brooks Cole, p 752

Pebriana RB, Rohman A, Lukitaningsih E, Sudjadi. (2017) Development of FTIR spectroscopy in combination with chemometrics for analysis of rat meat in beef sausage employing three lipid extraction systems. Int J Food Prop 20(sup2):1995–2005. https://doi.org/10.1080/10942912.2017.1361969

Pieszczek L, Czarnik-Matusewicz H, Daszykowski M (2018) Identification of ground meat species using near-infrared spectroscopy and class modeling techniques–aspects of optimization and validation using a one-class classification model. Meat Sci 139:15–24. https://doi.org/10.1016/j.meatsci.2018.01.009

Popping B, Buck N, Bánáti D, Brereton P, Gendel S, Hristozova N, Chaves SM, Saner S, Spink J, Willis C, Wunderlin D (2022) Food inauthenticity: authority activities, guidance for food operators, and mitigation tools. Compr Rev Food Sci Food Saf 21:4776

Yulirohyami Y, Maulidatunnisa V, Pusparani DP, Prasetyo B (2023) Identification of fat in pork using Fourier transform infrared spectrum and GC-MS. Indones J Chem Anal 6(2):187–194. https://doi.org/10.20885/ijca.vol6.iss2.art10

Rady A, Adedeji A (2018) Assessing different processed meats for adulterants using visible-near-infrared spectroscopy. Meat Sci 136:59–67. https://doi.org/10.1016/j.meatsci.2017.10.014

Rahayu WS, Rohman A, Martono S, Sudjadi S (2018a) Application of FTIR spectroscopy and chemometrics for halal authentication of beef meatball adulterated with dog meat. Indones J Chem 18:376–381. https://doi.org/10.22146/ijc.27159

Rahayu WS, Martono S, Sudjadi, Rohman A (2018b) The potential use of infrared spectroscopy and multivariate analysis for differentiation of beef meatball from dog meat for halal authentication analysis. J Adv Vet Anim Res 5:307–314. https://doi.org/10.5455/javar.2018.e281

Rahmania H, Rohman A (2015) The employment of FTIR spectroscopy in combination with chemometrics for analysis of rat meat in meatball formulation. Meat Sci 100:301–305

Ramos-Moreno L, Ruiz-Pérez F, Rodríguez-Castro E, Ramos J (2021) *Debaryomyces hansenii* is a real tool to improve a diversity of characteristics in sausages and dry-meat products. Microorganisms 9(7):1512. https://doi.org/10.3390/microorganisms9071512

Rani W, Man YB, Ismail A, Hashim P (2009) Fourier transform infrared (FTIR) spectroscopy differentiation of lard and other shortening in puff pastry. In: 3rd IMT-GT international symposium on Halal science and management (pp 37–42). KLIA Sepang, Malaysia

Rannou H, Downey G (1997) Discrimination of raw pork, chicken and Turkey meat by spectroscopy in the visible, near- and mid-infrared ranges. Anal Commun 34:401–404. https://doi.org/10.1039/A707694K

Rather JA, Akhter N, Ashraf QS, Mir SA, Makroo HA, Majid D, Barba FJ, Khaneghah AM, Dar BN (2022) A comprehensive review on gelatin: understanding impact of the sources, extraction methods, and modifications on potential packaging applications. Food Packag Shelf Life 34:100945. https://doi.org/10.1016/j.fpsl.2022.100945

Ray R, Prabhu A, Prasad D, Garlapati V k, Aminabhavi TM, Mani NK, Simal-Gandara J (2022) Based microfluidic devices for food adulterants: cost-effective technological monitoring systems. Food Chem 390:133173. https://doi.org/10.1016/J.FOODCHEM.2022.133173

Regenstein JM, Chaudry MM, Regenstein CE (2003) The kosher and halal food laws. Compr Rev Food Sci Food Saf 2:111–127. https://doi.org/10.1111/j.1541-4337.2003.tb00018.x

Restaino E, Fassio A, Cozzolino D (2011) Discrimination of meat patés according to the animal species by means of near infrared spectroscopy and chemometrics Discriminación de muestras de paté de carne según tipo de especie mediante el uso de la espectroscopia en el infrarrojo cercano y la quimiometria. CyTA-J Food 9(3):210–213. https://doi.org/10.1080/19476337.2010.512396

Rianti A, Novenia AE, Christopher A, Lestari D, Parassih EK (2018) Ketupat as traditional food of Indonesian culture. J Ethnic Foods 5(1):4–9. https://doi.org/10.1016/j.jef.2018.01.001

Rivas FP, Cayré ME, Campos CA, Castro MP (2018) Natural and artificial casings as bacteriocin carriers for the biopreservation of meats products. J Food Saf 38(1):e12419. https://doi.org/10.1111/jfs.12419

Rodriguez-Saona L, Aykas DP, Borba KR, Urtubia A (2020) Miniaturization of optical sensors and their potential for high-throughput screening of foods. Curr Opin Food Sci 31:136–150. https://doi.org/10.1016/j.cofs.2020.04.008

Rohman A (2018) The employment of Fourier transform infrared spectroscopy coupled with chemometrics techniques for traceability and authentication of meat and meat products. J Adv Vet Anim Res 6(1):9. https://doi.org/10.5455/javar.2019.f306

Rohman A, Che Man YB (2009) Analysis of cod-liver oil adulteration using Fourier transform infrared (FTIR) spectroscopy. J Am Oil Chem Soc 86:1149–1153. https://doi.org/10.1007/s11746-009-1453-9

Rohman A, Che Man YB (2010) FTIR spectroscopy combined with chemometrics for analysis of lard in the mixtures with body fats of lamb, cow, and chicken. Int Food Res J 17:519–526

Rohman A, Che Man YB (2011a) The optimization of FTIR spectroscopy combined with partial least square for analysis of animal fats in quartenary mixtures. J Spectrosc 25(3–4):169–176. https://doi.org/10.3233/SPE-2011-0500

Rohman A, Che Man YB (2011b) Analysis of lard in cream cosmetics formulations using ft-ir spectroscopy and chemometrics. Middle-East J Sci Res 7(5):726–732

Rohman A, Windarsih A (2020) The application of molecular spectroscopy in combination with chemometrics for halal authentication analysis: a review. Int J Mol Sci 21(14):5155. https://doi.org/10.3390/ijms21145155

Rohman A, Sismindari, Erwanto Y, Che Man YB (2011a) Analysis of pork adulteration in beef meatball using Fourier transform infrared (FTIR) spectroscopy. Meat Sci 88:91–95. https://doi.org/10.1016/j.meatsci.2010.12.007

Rohman A, Che Man YB, Hashim P, Ismail A (2011b) FTIR spectroscopy combined with chemometrics for analysis of lard adulteration in some vegetable oils Espectroscopia FTIR combinada con quimiometría para el análisis de adulteración con grasa de cerdo de aceites vegetales. Cyta-J Food 9(2):96–101. https://doi.org/10.1080/19476331003774639

Rohman A, Kuwat T, Retno S, Yuny E, Tridjoko W (2012) Fourier transform infrared spectroscopy applied for rapid analysis of lard in palm oil. Int Food Res J 19(3):1161

Rohman A, Gupitasari I, Purwanto KT, Rosman AS, Ahmad SAS, Yusof FM (2014) Quantification of lard in the mixture with olive oil in cream cosmetics based on FTIR spectra and chemometrics for halal authentication. Jurnal Teknologi (Sci Eng) 69(1). https://doi.org/10.11113/jt.v69.2062

Rohman A, Arsanti L, Erwanto Y, Pranoto Y (2016) The use of vibrational spectroscopy and chemometrics in the analysis of pig derivatives for halal authentication. Int Food Res J 23(5):1839–1848

Rohman A, Himawati A, Kuwat Triyana S, Fatimah S (2017) Identification of pork in beef meatballs using Fourier transform infrared spectrophotometry and real-time polymerase chain reaction. Int J Food Prop 2017(20):654–661. https://doi.org/10.1080/10942912.2016.1174940

Salamah N, Jufri SL, Susanti H, Jaswir I (2023) Analysis of gelatin on soft candy using a combination of Fourier Transform Infrared Spectroscopy (FTIR) with chemometrics for halal authentication. Indones J Halal Res 5(2):90–98. https://doi.org/10.15575/ijhar.v5i2.25682

Salleh NAM, Hassan MS, Jumal J, Harun FW, Jaafar MZ (2018) Differentiation of edible fats from selected sources after heating treatments using Fourier transform infrared spectroscopy (FTIR) and multivariate analysis. In: Proceedings of the AIP Conference Proceedings, Melaka, Malaysia, 7–8 November 2017, vol 1972. American Institute of Physics Inc, College Park. https://doi.org/10.1063/1.5041236

Saputra I, Jaswir I, Akmeliawati R (2018) Identification of pig adulterant in mixture of fat samples and selected foods based on FTIR-PCA wavelength biomarker profile. Int J Adv Sci Eng Inf Technol 8(6):2341–2348

Sari TNI, Guntarti A (2018) Wild boar fat analysis in beef sausage using FTIR method (Fourier transform infrared) combined with chemometrics. Jurnal Kedokteran dan Kesehatan. Indonesia 9(1):16–23. https://doi.org/10.20885/JKKI.Vol9.Iss1.art4

Schmutzler M, Beganovic A, Böhler G, Huck CW (2015) Methods for detection of pork adulteration in veal product based on FT-NIR spectroscopy for laboratory, industrial and on-site analysis. Food Control 57:258–267. https://doi.org/10.1016/j.foodcont.2015.04.019

Shawky E, Nahar L, Nassief SM, Sarker SD, Ibrahim RS (2024) Dairy products authentication with biomarkers: a comprehensive critical review. Trends Food Sci Technol 147:104445. https://doi.org/10.1016/j.tifs.2024.104445

Sheikh M, Islam T (2018) Islam, alcohol, and identity: towards a critical Muslim studies approach. ReOrient 3(2):185–211. https://doi.org/10.13169/reorient.3.2.0185

Siddiqui MA, Md Khir MH, Ullah Z, Al-Hasan M'a, Saboor A, Magsi SA (2023) Infrared spectroscopy-based chemometric analysis for lard differentiation in meat samples. Comput Mater Continua 75(2):10.32604/cmc.2023.034164

Silva LCR, Folli GS, Santos LP, Barros IHAS, Oliveira BG, Borghi FT, dos Santos FD, Filgueiras PR, Romão W (2020) Quantification of beef, pork, and chicken in ground meat using a portable NIR spectrometer. Vib Spectrosc 111:103158. https://doi.org/10.1016/j.vibspec.2020.103158

Silverstein RM, Bassler GC, Morrill TC (1981) Spectrometric identification of organic compounds, 4th edn. Wiley, New York

Sim SF, Chai MXL, Kimura ALJ (2018) Prediction of lard in palm olein oil using simple linear regression (SLR), multiple linear regression (MLR), and partial least squares regression (PLSR) based on Fourier-transform infrared (FTIR). J Chem 2018(1):7182801. https://doi.org/10.1155/2018/7182801

Siska S, Jumadil MI, Abdullah S, Ramadon D, Mun'im A (2023) ATR-FTIR and chemometric method for the detection of pig-based derivatives in food products-a review. Int Food Res J 30(2):9. https://doi.org/10.47836/ifrj.30.2.01

Soon-Sinclair JM, Ha TM, Vanany I, Limon MR, Sirichokchatchawan W, Wahab IRA, Hamdan RH, Jamaludin MH (2024) Consumers' perceptions of food fraud in selected southeast Asian countries: a cross sectional study. Food Secur 16(1):65–77. https://doi.org/10.1007/s12571-023-01406-z

Stuart BH (2004) Infrared spectroscopy: fundamentals and applications. Wiley

Stuart BH (2007) Analytical techniques in materials conservation. Wiley

Supandi S, Septiana AD, Kusumadewi N, Fatmawati S (2024) Halal authentication: Fourier transform infrared spectroscopy and multivariate calibration application for pork gelatin analysis in gummy candy. Food Res 8(5):44–48. https://doi.org/10.26656/fr.2017.8(5).573

Suparman S, Rahayu WS, Sundhani E, Saputri SD (2015) The use of Fourier transform infrared spectroscopy (FTIR) and gas chromatography-mass spectroscopy (GCMS) for halal authentication in imported chocolate with various variants. J Food Pharm Sci 3(1):6–11. https://doi.org/10.14499/jfps

Surowiec I, Fraser PD, Patel R, Halket J, Bramley PM (2011) Metabolomic approach for the detection of mechanically recovered meat in food products. Food Chem 125(4):1468–1475. https://doi.org/10.1016/j.foodchem.2010.10.064

Syabani MW, Iswahyuni I, Warmiati W, Prayitno KA, Saraswati H, Hernandha RFH (2023) Unveiling the signature of halal leather: a comparative study of surface morphology, functional groups and thermal characteristics. Indonesian J Halal Res 5(2):67–76. https://doi.org/10.15575/ijhar.v5i2.25702

Syahariza ZA (2006) Detection of lard in selected food model systems using Fourier transform infrared spectroscopy (Doctoral dissertation, Universiti Putra Malaysia)

Syahariza Z a, Che Man YB, Selamat J, Bakar J (2005) Detection of lard adulteration in cake formulation by Fourier transform infrared (FTIR) spectroscopy. Food Chem 92(2):365–371. https://doi.org/10.1016/j.foodchem.2004.10.039

Syofyan S, Fadillah R, Zehanna D, Hamidi D, Syafri S (2025) Soxhlet extraction and FT-IR spectroscopy coupled to chemometrics: authenticating Beef, Pork and Wild boar Rendang. Curr Nutr Food Sci. https://doi.org/10.2174/0115734013341980241204180210

Thanasi V, Caldeira I, Santos L, Ricardo-da-Silva JM, Catarino S (2024) Simultaneous determination of ethanol and methanol in wines using FTIR and PLS regression. Foods 13(18):2975. https://doi.org/10.3390/foods13182975

Totaro MP, Squeo G, De Angelis D, Pasqualone A, Caponio F, Summo C (2023) Application of NIR spectroscopy coupled with DD-SIMCA class modelling for the authentication of pork meat. J Food Compos Anal 118:105211. https://doi.org/10.1016/j.jfca.2023.105211

Upadhyay N, Jaiswal P, Jha SN (2018) Application of attenuated total reflectance Fourier transform infrared spectroscopy (ATR–FTIR) in MIR range coupled with chemometrics for detection of pig body fat in pure ghee (heat clarified milk fat). J Mol Struct 1153:275–281. https://doi.org/10.1016/j.molstruc.2017.09.116

Uríčková V, Sádecká J (2015) Determination of geographical origin of alcoholic beverages using ultraviolet, visible and infrared spectroscopy: a review. Spectrochim Acta A Mol Biomol Spectrosc 148:131–137. https://doi.org/10.1016/j.saa.2015.03.111

USFDA (2004) Chapter VI: Cosmetics. In: The federal food, drug and cosmetic act (FD&C act). U.S. Food and Drug Administration, Silver Spring

Utami PI, Ryandita I, Sundhani E (2018) Fourier transform infrared spectroscopy for identification of pig oil in imported instant noodle spice oil. In: The 8th university research colloquium 2018. Universitas Muhammadiyah Purwokerto, Purwokerto, pp 686–694

Van Mierlo K, Baert L, Bracquené E, De Tavernier J, Geeraerd A (2022) Moving from pork to soy-based meat substitutes: evaluating environmental impacts in relation to nutritional values. Future Foods 5:100135. https://doi.org/10.1016/j.fufo.2022.100135

Vichasilp C, Poungchompu O (2014) Feasibility of detecting pork adulteration in halal meatballs using near infrared spectroscopy (NIR). Chiang Mai Univ J Nat Sci 13(1):497–507. https://doi.org/10.12982/CMUJNS.2014.0052

Wang L, Sun D-W, Pu H, Cheng J-H (2017) Quality analysis, classification, and authentication of liquid foods by near-infrared spectroscopy: a review of recent research developments. Crit Rev Food Sci Nutr 57(7):1524–1538. https://doi.org/10.1080/10408398.2015.1115954

Wang M, Cui J, Wang Y, Yang L, Jia Z, Gao C, Zhang H (2022a) Microfluidic paper-based analytical devices for the determination of food contaminants: developments and applications. J Agric Food Chem 70(27):8188–8206. https://doi.org/10.1021/acs.jafc.2c02366

Wang F, Chandler PD, Zeleznik OA, Wu K, Wu Y, Yin K, Giovannucci EL (2022b) Plasma metabolite profiles of red meat, poultry, and fish consumption, and their associations with colorectal cancer risk. Nutrients 14(5):978. https://doi.org/10.3390/nu14050978

Waskitho D, Lukitaningsih E, Sudjadi, Rohman A (2016) Analysis of lard in lipstick formulation using FTIR spectroscopy and multivariate calibration: a comparison of three extraction methods. J Oleo Sci 65(10):815–824. https://doi.org/10.5650/jos.ess15294

Wiedemair V, De Biasio M, Leitner R, Balthasar D, Huck CW (2018) Application of design of experiment for detection of meat fraud with a portable near-infrared spectrometer. Curr Anal Chem 14(1):58–67. https://doi.org/10.2174/1573411013666170207121113

Windarsih A, Irnawati AR (2020) Application of FTIR-ATR spectroscopy and chemometrics for the detection and quantification of lard oil in bovine milk fat. Food Res 4(5):1732–1738. https://doi.org/10.26656/fr.2017.4(5).087

Windarsih A, Indrianingsih AW, Apriyana W, Rohman A (2023) Rapid detection of pork oil adulteration in snakehead fish oil using FTIR-ATR spectroscopy and chemometrics for halal authentication. Chem Pap 77(5):2859–2870. https://doi.org/10.1007/s11696-023-02671-0

Windarsih A, Jatmiko TH, Anggraeni AS, Rahmawati L (2024) Machine learning-assisted FT-IR spectroscopy for identification of pork oil adulteration in tuna fish oil. Vib Spectrosc 134:103715. https://doi.org/10.1016/j.vibspec.2024.103715

Wirnawati W, Lestari D, Rohman A, Andayani R, Hamidi D (2023) Analysis of adulteration dog meat in beef sausages using FTIR spectroscopy combined with chemometrics. Res Sq. https://doi.org/10.21203/rs.3.rs-3195476/v1

Wu Q, Kamruzzaman M (2024) Advancements in nanozyme-enhanced lateral flow assay platforms for precision in food authentication. Trends Food Sci Technol 147:104472. https://doi.org/10.1016/j.tifs.2024.104472

Xu L, Cai CB, Cui HF, Ye ZH, Yu XP (2012) Rapid discrimination of pork in Halal and non-Halal Chinese ham sausages by Fourier transform infrared (FTIR) spectroscopy and chemometrics. Meat Sci 92(4):506–510. https://doi.org/10.1016/j.meatsci.2012.05.019

Yang H, Irudayaraj J (2001a) Comparison of near-infrared, Fourier transform-infrared, and Fourier transform-Raman methods for determining olive pomace oil adulteration in extra virgin olive oil. J Am Oil Chem Soc 78:889–895. https://doi.org/10.1007/s11746-001-0360-6

Yang H, Irudayaraj J (2001b) Characterization of beef and pork using Fourier-transform infrared photoacoustic spectroscopy. LWT 34(6):402–409. https://doi.org/10.1006/fstl.2001.0778

Yang H, Irudayaraj J, Paradkar MM (2005) Discriminant analysis of edible oils and fats by FTIR, FT-NIR and FT-Raman spectroscopy. Food Chem 93(1):25–32. https://doi.org/10.1016/j.foodchem.2004.08.039

Yang L, Ting W, Liu Y, Zou J, Huang Y, Sarath BV, Lin L (2018) Rapid identification of pork adulterated in the beef and mutton by infrared spectroscopy. J Spectrosc 2018(1):2413874. https://doi.org/10.1155/2018/2413874

Zilhadia Z, Kusumaningrum F, Betha OS, Supandi S (2018) Differentiation of bovine and porcine gelatin extracted from vitamin C gummy by combination method of Fourier transform infrared (FTIR) and principal component analysis (PCA). Pharm Sci Res 5(2):90–96. https://doi.org/10.7454/psr.v5i2.4013

Zilhadia Z, Anggraeni Y, Apriyanti YF, Mustafidah M, Jaswir I (2024) Analysis of lard in cod liver oil emulsion using FTIR spectroscopy combined principal component analysis. Food Res 8(3):424–431. https://doi.org/10.26656/fr.2017.8(3).329

Zwinkels J (2015) Light, electromagnetic spectrum. In: Encyclopedia of color science and technology, vol 8071, pp 1–8. https://doi.org/10.1007/978-3-642-27851-8_204-1

Chapter 5
Halal Meat, Fat, Oil, and Cosmetics Authentication

5.1 Meat Authentication

Meat and meat products from different sources, i.e., animals, birds, and fish, are an essential part of the human diet, providing proteins, vitamins, and minerals. Human sustainability and growth require certain amounts of these on a daily basis. Consumer mindset plays a role in the selection and use of meat and products. Consumers are influenced by religion, social norms, and health-related issues when selecting meat. Consumers prefer authentic and non-adulterated products. Awareness of meat and ingredients is a key consideration in the selection of items. Sometimes, the ingredients and source of meat are not clearly mentioned or can't be readily verified; in these conditions, authentication is very necessary.

In the authentication of meat, five things are mostly considered: first is the species of meat, second is production/slaughter, third is processing methods and conditions, fourth is the origin of meat species, and fifth is ingredients used. Standard techniques are present for each step, and for every condition, authentication is performed differently depending upon the needs of consumers. Authentication can be either performed by the producer or authentication bodies. Meat authentication encompasses assessing quality, labeling requirements, and technical specifications.

5.1.1 Meat Geography Authentication

Meat components and quality are directly influenced by food, water, and the environment raised by animals. The composition of meat is directly linked with the feed of the animals. The markers mentioned above provide details about the origin and dietary background of animals (Fig. 5.1). Certain feedstuffs are commonly used in particular geographical areas, so authentication of the geographical origin of a

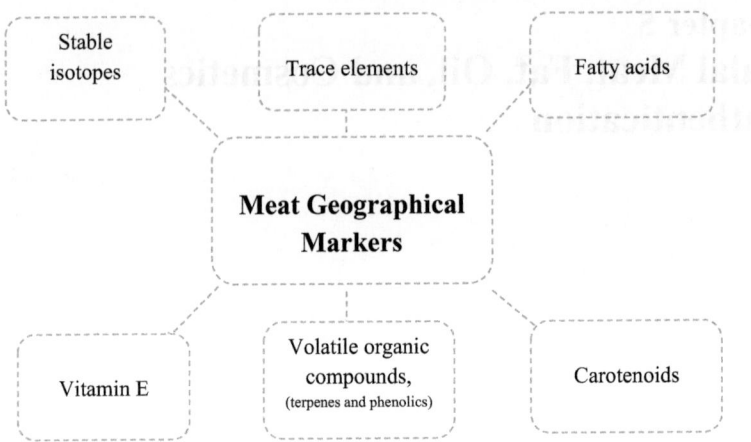

Fig. 5.1 Origin-based markers for meat

species is possible (Schmidt et al. 2005) and is influenced by regional climatic conditions or underlying geology (Capo et al. 1998). It is well established that the stable isotope composition of bioelements in animal tissue is influenced by the composition of the diet consumed by the animal (DeNiro and Epstein 1978).

5.1.2 Meat Origin Authentication

5.1.2.1 Sex

Hormone analysis can confirm the sex of meat. The level of hormones can vary in individuals. According to Zeleny and Schimmel (2002), hormone analysis does not provide authentic information about the sex of animals, so chromatography (GC-MS and HPLC-MS/MS) is performed to authenticate the sex of animals (Hartwig et al. 1997; Draisci et al. 2000) and enzyme-linked immunosorbent assays (ELISA) (Simontacchi et al. 1999).

5.1.2.2 Breed Type Identification

Authentication of breed determines the quality and authenticity of meat. Breed authentication is done using DNA-based analytical techniques. Other analytical methods can be used alone or in combination. Statistical model converts raw data obtained from analysis into concrete results. Dalvit et al. used PCR with a Bayesian statistical model to authenticate and identify Italian cow meat (Dalvit et al. 2008). Another study used DNA for the identification of Holstein and Japanese cattle breeds (Sasazaki et al. 2004).

5.1.2.3 Animal Feed Type

Tracing of animal feed is sometimes used to authenticate meat. Components of animal feed can be traced to meat, blood, and milk. Sometimes, the metabolites of feed are traced. Different feeds show different constituents in intact or metabolized form, i.e., pasture hay and maize. Animal meat and fat contain these constituents in different forms and quantities. "The carotenoids, xanthophylls, and carotenes are much more abundant in the pasture when compared to hay and concentrate. HPLC can be used to measure carotenoids in sheep's blood" (Prache et al. 2003). "However, intrinsic factors such as breed, gender, lactation, and rumen environment also affect the carotenoid content" (Dunne et al. 2009). "The composition of fatty acids in meat fat is also dependent on an animal's diet. GC studies describe a higher ratio of polyunsaturated fatty acids to saturated fatty acids" (Duckett et al. 1993).

5.1.2.4 Slaughter Age

Slaughter age is considered in Islam only for *Qurbani* (sacrifice) of animals on Eid (holy occasion). Slaughter age is important only for non-Islamic reasons. Meat from younger animals is more desired due to its juiciness, tenderness, and abundance of nutrients (Kawęcka and Pasternak 2022). Veal is considered more valuable than beef, and lamb is more valuable than mutton. The concept of veal and lamb compared to beef and mutton varies in different societies. This slightest difference can complicate analytical authentication. In the countries of the European Union, veal must come from animals aged 12 months or older. This can't be verified using analytical methods. Feed type can be used to confirm by analytical methods. If the veal label shows whether the animal is milk-fed or grain-fed, the milk and grain-fed animals' meat differ in taste and quality, and analytical testing can confirm specific constituents from the feed that are found in the meat (Ballin 2010).

5.1.3 Organic Versus Conventional Meat

Organic meat refers to meat that comes from animals that are fed organic feed, grown in a natural environment, and raised in accordance with organic farming standards. Organic farming standards include proper treatment of animals and minimal use of hormones and antibiotics. Consumers' interest in organic meat has been increasing globally due to factors such as health concerns, environmental awareness, and animal welfare issues. The use of antibiotics and drugs in organic meat farming is strictly controlled as to when and why they should be used. Spectroscopic methods can be used to detect tetracycline in chicken and beef, which can help verify whether the animal complies with the specified organic rules. Fat distribution in animals can be used to identify organically and conventionally raised animals. A GC study of fatty acid methyl esters showed an increase in polyunsaturated fatty acids in organically produced lamb (Angood et al. 2008) and broiler (Kim et al. 2009).

5.1.4 Identification of Meat Substitution

Meat can be adulterated either by substituting tissue or species. Identification of species is mostly done using protein analysis. Recent advancements have led to the determination of species through DNA rather than proteins. Due to the thermal stability of DNA, it can be used alone to identify species. Every tissue provides the same information, and DNA is available in most tissues (Lockley and Bardsley 2000). "Currently, real-time PCR is generally the method of choice and a number of qualitative" (Brodmann and Moor 2003; Fumière et al. 2006) and quantitative (Chisholm et al. 2005; López-Andreo et al. 2006) species determination methods have been published.

There are a few types of PCR that can be used to identify and differentiate species of meat. Multiplex PCR can be used to identify multiple species simultaneously. Restriction fragment length polymorphism is another type of PCR that utilizes enzymatic digestion of DNA fragments and then amplifies DNA fragments (Verkaar et al. 2002). RAPD (Random Amplified Polymorphic DNA) is used to identify unknown species. It can be determined without prior DNA data, and analysis takes advantage of short arbitrary primers. It can differentiate domestic as well as wild animal species (Calvo et al. 2001).

Characterization of animal species is done by comparing it with the data available in the database. There are many extensive databases available that contain detailed information on species. One of the biggest databases is the NIH database (http://www.ncbi.nlm.nih.gov/), which includes information on the DNA sequence of animal species. It can be used to identify animal species without having references available. Sequenced PCR amplicons and identification through database comparison have several cases established the identity of the species (Bartlett and Davidson 1992; Forrest and Carnegie 1994; Iijima et al. 2006). The incorporation of collagen and offal is an adulteration. Collagen and elastin can be authenticated by confirming the percentage of hydroxyproline, which is 8% in collagen and 1% in elastin (Etherington and Sims 1981).

5.1.5 Vegetable Protein and Animal Protein

Soy protein, which is cheaper than some animal proteins, can be used to mix in animal proteins. Detection of soy protein can be easily done with ELISA (González-Córdova et al. 1998). Certain histidine peptides are only present in animals, i.e., carnosine, anserine, and balenine (Aristoy and Toldrá 2004). Peptides can be tissue and species-specific. Carnosine and anserine are present in specific amounts in pork, beef, mutton, and chicken (Aristoy and Toldrá 2004). Certain muscle proteins can be separated and quantified using isoelectric focusing and can be used to identify similar fish species (Berrini et al. 2006).

5.2 Oil and Fat Authentication

Fats and oils are composed of triacylglycerols (TAGs), diacylglycerols (DG's), free fatty acids (FFAs), phospholipids, and minor other constituents. Triacylglycerols and FFAs are the most important ones (Buchgraber et al. 2004). TAGs usually differ in total carbon number, degree of unsaturation, and the position and structure of double bonds within each FA. The exact position of the three FAs on the glycerol backbone determines the TAG molecule's region- and stereo-specificity. Because of the many FA combinations on the glycerol backbone, each oil or fat might include multiple TAGs. Animal fats are quite complex and may contain a variety of FAs. Ruminant lipids contain extra FAs due to ruminal microbial degradation. Milk fat contains more than 400 distinct FAs.

The human diet contains a major portion of oils and fats (vegetal or animal). Many food products include a large proportion of these, i.e., cocoa butter, margarine, dressings, cooking oil, etc. The health benefits of these products depend upon the source and quality of the products. Multiple parameters need to be properly addressed to attain the quality of fats and oils.

Physical and chemical parameters like boiling point and oxidation value determine the overall quality of fats and oils. These parameters can be traced to the authenticity of the fats and oils (Bell and Gillatt 1994).

The solid fat index (SFI) is a key indicator of the proportion of triglycerides in fats in solid form and is detected with the help of dilatometry. The dilatometry analysis gives information about changes in specific volumes of fat as a function of temperature. Dilatometry results are strictly dependent upon the method followed. NMR can be used to detect solid fat content (SFC), and a rapid method based on MIR and PLS for the determination of SFI is also possible.

5.2.1 Animal and Vegetable Fat

Vegetable fat adulteration in animal fat can be confirmed by the presence of phytosterols, i.e., stigmasterol and β-sitosterol. These facts are detected using GC-MS (Nair et al. 2006; Szűcs et al. 2006) and atmospheric pressure photoionization (APPI) LC-MS/MS (Lembcke et al. 2005) in various matrices. It is of no importance to detect the ratio of these fats. Quantitative analysis of particular fats can be used to detect adulteration of animal and vegetable species (Precht 1992).

5.2.2 Lard and Pork Authentication

NMR spectroscopy is one of the fundamental techniques to identify and elucidate the structure of small and large molecules utilizing isotopic ratios (Ashbrook and Hodgkinson 2018). NMR can provide fingerprint profiles of compounds that can be used to authenticate Halal foods and components (Petrakis et al. 2015). However, NMR spectroscopy has limited sensitivity but is still capable of identifying components of foods, cosmetics, and pharmaceuticals. NMR can analyze multi-component samples utilizing small amounts of sample and solvent. Reproducibility and robustness are key to this technique (Windarsih et al. 2019). NMR spectroscopy can analyze samples in all three phases (solid, liquid, and gas) (Emwas 2015). Proton-NMR (H^1-NMR) has been used to identify lard adulteration in butter. There are characteristic peaks that appear in the region of 2.60–2.84 ppm; the lard peak occurs at 2.63 ppm. This peak seems to be due to double allylic methylene protons (Fadzillah et al. 2017). H^1-NMR can be applied to crude samples efficiently (Awin et al. 2016). Quantitative NMR has been utilized to differentiate chicken, beef, and pork meats. The difference in peaks of NMR is due to the change in the amino acid composition of these meats (Kim et al. 2010). Figure 5.2 shows the advantages and disadvantages of NMR spectroscopy. The high cost and complexity of instruments limit the use of this technique. Skilled personnel and large space are required to operate NMR instruments (Emwas 2015).

NMR is an analytical tool that provides a structural arrangement of atoms and orientation that can be used to identify known and unknown compounds. NMR utilizes radio frequency to assess the nuclear magnetic field in the presence of an external magnetic field. Only odd atomic number nuclei can be detected with NMR,

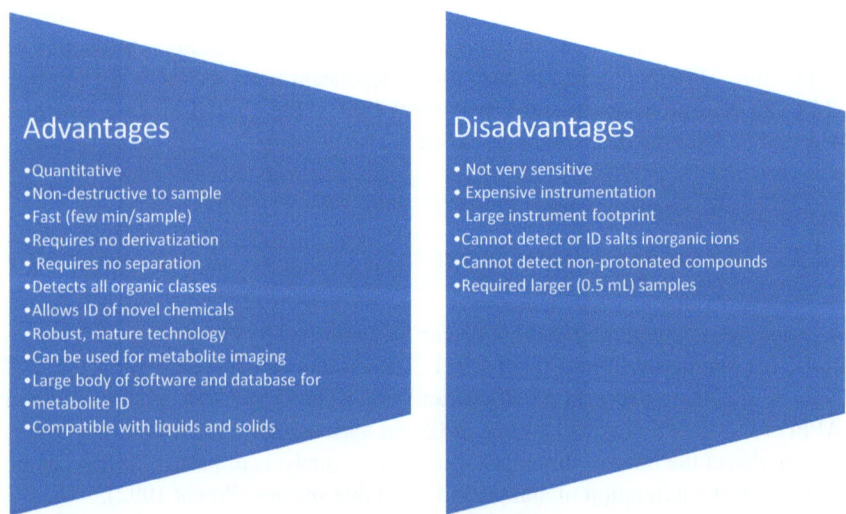

Fig. 5.2 NMR spectroscopy advantages and disadvantages

i.e., C^{13}, H^1. NMR requires larger samples for analysis, and deuterated solvents are used to analyze them. Analysis of foods and components, i.e., oils, fats, and lipids, is possible with this technique, and it can detect origin, quality, and other parameters. Fatty acid profiles of fats and oils can be obtained, which provides information about nutritional value and quality (Tang et al. 2019). In general, NMR can identify origin-related issues (Masoum et al. 2007), and this technique was used to determine the origin of salmon samples. Also, compositional analysis of natural sugar-based products, i.e., honey, makes it possible to authenticate the origin of honey to evaluate its authenticity. Constitutional analysis consisting of different components, including water, fats, and proteins is possible for different kinds of foods. NMR based upon resolution power has different types such as low resolution, high resolution H^1 and C^{13} NMR.

NMR can be used to check solid fats and liquid oils in samples by analyzing changes that occur when temperature is changed. H^1 NMR and C^{13} NMR can be used to check saturation and unsaturation and identify types of FAs present that can ultimately determine the type of fats and oils. NMR can distinguish between the fatty acid profiles of horses and beef. Principle component analysis can be used in very specific NMR regions to identify beef and horse meat. Bis-allylic, olefinic, and the terminal CH_3 peaks served as three key distinguishing signals utilized to differentiate between horse meat and beef (Jakes et al. 2015).

Fang et al. (2013) conducted a quantitative analysis of processed pork. They found that processed samples of pork from Italy contain high amounts of polyunsaturated fatty acid chains and the majority of oleyl chains. H^1 NMR can identify adulteration of pork and beef fat in canola oil. TAG profiles can be detected not only by HPLC but also by NMR as a marker for Halal/kosher issues.

H^1 NMR spectroscopy can be used for adulteration detection of lard by the presence of peaks at 2.60–2.84 ppm. Peak attributed to double-allylic methylene protons of polyunsaturated FAs (Fadzillah et al. 2017) and linoleic acid signals at δ 2.63 ppm in seven peaks show the presence of lard in butter. High-resolution NMR can be used to identify specific markers composed of fats in lard for Halal authentication. NMR is very limited in the applicability of Halal food authentication due to its high cost and low sensitivity compared to chromatographic methods and mass spectrometry. At this point, to create a wider range of applications for this technique, further research and investigation efforts are needed to improve NMR in food identification (Mortas et al. 2022).

5.3 Cosmetics Authentication

Halal cosmetics are products composed of Halal ingredients and prepared via Halal processes; cosmetics are intended for different applications depending upon the products, i.e., lipsticks and powder. The applications of cosmetics can vary depending on time and period; sometimes, they are washed immediately, while at other times, they are applied for longer periods. Cosmetics contain different constituents;

each ingredient has unique properties and functions, i.e., water is used to dissolve ingredients. Oils can also be used to dissolve hydrophobic ingredients. Other commonly used ingredients are surfactants, polymers, organic solvents, colorants, proteins, vitamins, natural extracts, preservatives, and antioxidants, among others (Iwata and Shimada 2012). When making cosmetic products, the ingredients and their origins are very important. For Halal cosmetics, where adherence to religious rules is essential, the source of the ingredients significantly impacts the quality of the product. Not only the government but companies need to make sure all ingredients in Halal cosmetics are safe. They work closely with suppliers to use only Halal-certified ingredients. It is not just about rules. It is about making sure products are safe and good for people who use them (Mathew 2014). Halal cosmetics are considered of higher quality and safer compared to traditional cosmetics. This is the result of strict guidelines and certification processes in the manufacturing of Halal cosmetics (Sugibayashi et al. 2019). Manufacturing of Halal cosmetics requires close collaboration and cautious monitoring of ingredient suppliers, manufacturers, storage houses, and distributors. Each department, from supplier to storage, should be Halal certified. Every raw material, actives, or excipients must come from certified Halal sources. The safety and Halal integrity of these ingredients should be thoroughly verified.

5.3.1 Halal (Permitted) Cosmetic Ingredients

Halal cosmetics' ingredients can be from natural or synthetic sources. The natural sources include plants, animals, birds, fish, microbes, etc. Natural sources of plant and fish origin are usually preferred for Halal cosmetics. Safe synthetic materials can be used as ingredients in Halal cosmetics. Increasingly, cosmetic companies are moving towards plant-based and synthetic alternatives instead of relying on

Table 5.1 Halal cosmetic ingredients

Category	Examples	References
Skin-whitening agents	4-MSK, ferulic acid, tranexamic acid, vitamin B3, vitamin C	Linster and Van Schaftingen (2007), Kumar and Pruthi (2014), Li et al. (2015) and Li et al. (2016)
Anti-aging agents	Capsanthin, genistein, glycyrrhizin, lutein, vitamin B3	Brieskorn and Sax (1970), Kim and DellaPenna (2006), Yamano and Ito (2007) and Xiong et al. (2015)
Colorants	Carotene, paprika, turmeric	Iwata et al. (2013)
Thickeners	Carboxymethyl cellulose, carnauba wax, carrageenan, petrolatum	Battarjee et al. (1999), Shui et al. (2017), Yu et al. (2017) and Zia et al. (2017)
Solvents	Avocado oil, corn oil, cottonseed oil, liquid paraffin, polyethylene glycol, sesame oil, water	Qin and Zhong (2016)

Table 5.2 Haram cosmetic ingredients

Haram ingredient	Examples	References
Restricted chemicals	Chlorofluorocarbon propellants, chloroform, halogenated salicylanilides, hexachlorophene, mercury compounds, methylene chloride, prohibited cattle materials, vinyl chloride, zirconium-containing complexes	https://www.fda.gov/cosmetics/cosmetics-laws-regulations/prohibited-restricted-ingredients-cosmetics
Insect derived	Carmine dye (cochineal), crimson dye (from kermes vermilio), laccaic acid, beeswax	Hepburn and Hepburn (1986) and Iwata et al. (2013)
Human derived	Amniotic fluid, placenta	Yusuf and Yajid (2017)
Porcine derived	Amniotic fluid, gelatin, lard	Kim et al. (2011), Lukitaningsih et al. (2012) and Yusuf and Yajid (2017)

ingredients derived from animals. Using critical cosmetic ingredients may hinder the Halal certification process. Verifying the source of critical ingredients is a lengthy and costly process. The cosmetic ingredients listed below are classified as Halal under conditions that they are not contaminated with haram and are produced via the Halal process. Table 5.1 lists the Halal cosmetic ingredients, while Table 5.2 lists the haram cosmetic ingredients.

5.3.2 Critical Cosmetic Ingredients

Critical cosmetic ingredients are those ingredients that are derived from unauthentic sources, non-Halal slaughtered animals, contaminated materials, and adulterated materials, i.e., if gelatin or collagen is used, its source must be confirmed. Ingredients that are processed in such a manner that haram ingredients become part of it. For critical ingredients, proper authentication and certification are often required. The use of alcohol (ethanol) is prohibited in food but allowed as cosmetic ingredients. If alcohol is obtained from natural fermentation in the presence of oxygen, it is permitted, but if obtained from liquor, it is not permitted. Table 5.3 provides a representative list of ingredients classified as critical.

5.3.3 Detection of Haram Ingredients in Cosmetics

Halal authentication of cosmetics is also like Halal authentication of food. Both required authentication of ingredients, processes, and products. Analyses and authentication of cosmetics and their ingredients require analytical tools that are

Table 5.3 Types of critical ingredients and their sources

Category	Ingredients	Critical source/process/contamination	References
Actives	Allantoin	Animal urine	Cativiela et al. (2003)
Actives	Caffeic acid	Microbes/bee propolis/Halal if plant-derived	Lin and Yan (2012 and Zhang et al. (2014)
Actives	Collagen	Porcine-derived/ human-derived/ Halal marine-derived	Avila Rodríguez et al. (2018)
Actives	Hyaluronic acid	Haram animal tissues	Sze et al. (2016)
Actives	Mequinol	Methanol	Couteau and Coiffard (2016)
Actives	Oligopeptides	Microorganisms/haram animals	Schagen (2017)
Actives	Urea	Unknown animals	Yusuf and Yajid (2017)
Actives	Vitamin E	Non-Halal processes (i.e., use of lipase or unspecified origin of precursor materials)	Netscher (2007)
Thickeners	Gelatin	Porcine-derived/Halal if derived from fish	Bagal-Kestwal et al. (2019)
Thickeners	Xanthan gum	Haram fermenting bacterium/Halal if natural aerobic fermentation	Lopes et al. (2015)
Oils	Linoleic acid/ Linolenic acid	Unspecified animals/Halal if plant-derived	Imanaka et al. (1999)
Oils	Oleic acid	Porcine-derived	Nagai and Bloch (1965)
Oils	Palm kernel oil	Unspecified animals	Rahman et al. (2003)
Waxes	Lanolin alcohol	Non-Halal slaughtered animals/ Halal if obtained from living animals	Schlossman and Mccarthy (1978)
Waxes	Stearyl alcohol	Stearic acid of unspecified animal origin	Zhen et al. (2015)
Solvents	Glycerin/glycerol	Porcine-derived	DFG (2015)
Solvents	Propylene glycol	Glycerol of unspecified animal origin	Seretis and Tsiakaras (2016) and Sugibayashi et al. (2019)

used in food analyses. Authentication of cosmetic ingredients involves a multitude of methods and processes (Kim et al. 2018).

Halal cosmetics must contain Halal ingredients, and all the processes of production must meet Islamic Sharia law. It also includes maintaining the purity and quality of products. Certain products may interfere with Islamic rituals, i.e., *Wudu* and *gusul* (washing of the whole body along with mouth and nose); such products may be evaluated carefully (Fatima et al. 2023).

Some common haram ingredients in cosmetics include casein, oleic acid, lauric acid, gelatin, collagen, carmine, and alcohol. These ingredients can be derived from animals or plants, and their permissibility is a subject of controversy within the Islamic community. Collagen obtained from pork, non-Halal animals, and animals slaughtered not according to Islam is considered haram, and collagen obtained from poultry and fish is considered Halal. Chromatographic methods, FTIR, and PCR can be used to identify gelatin. Casein is haram if obtained from haram animals. Casein can be identified with FTIR and LC-MS. Additionally, some coloring agents and substances derived from bovines may also be considered haram by certain competent Halal certification bodies. It is recommended that you opt for Halal-certified products to ensure that they adhere to Islamic standards and do not contain any haram ingredients.

"To identify haram ingredients in cosmetics, it is crucial to conduct thorough research and due diligence. This involves checking the ingredient list of each product and verifying their sources. It is important to note that some ingredients may be derived from both Halal and haram sources, so it is essential to confirm their origin. Haram fats commonly used in cosmetics include lard, and FAs derived from pig. These ingredients can be identified using analytical methods such as FTIR spectroscopy, chromatography, and PCR. These methods have been successfully developed and validated for the detection of non-Halal components in cosmetics. Identification of different components can vary widely depending upon the analytical method used and the deterioration of components due to thermal, physical, and chemical modification in manufacturing and storage. The stability of DNA gives an opportunity for identification even after modification. Biological materials used in cosmetics are not highly purified, so the presence of DNA is mostly possible. PCR is a rapid, accurate, and highly sensitive method that can selectively amplify a small amount of target DNA. NMR and Raman spectroscopy are advanced analytical techniques that can be used to identify haram fats in cosmetics. These techniques are based on the detection of specific chemical bonds and molecular vibrations, which can provide information about the composition and structure of the sample. For example, NMR spectroscopy has been used to identify lard in lipstick formulations and processed edible oils and fats" (Fadzillah et al. 2017). Raman spectroscopy has also been used to classify and quantify animal fats in mixtures, including lard, beef tallow, chicken fat, and duck oil (Lee et al. 2018).

Haram proteins used in cosmetics include those derived from forbidden animals, such as pigs, and human body parts or blood. These proteins can be found in various cosmetic products, including creams, lotions, and mask packs. They can be identified using analytical methods such as NMR spectroscopy, Raman spectroscopy, and real-time PCR. A product's Halal status cannot be decided by only the presence or absence of Halal and haram ingredients. But it must also not hinder any Islamic rituals performed by Muslims. Products should be developed in such a way that their consistency and performance adhere to Islamic values and traditions.

References

Angood K et al (2008) A comparison of organic and conventionally-produced lamb purchased from three major UK supermarkets: Price, eating quality and fatty acid composition. Meat Sci 78(3):176–184

Aristoy MC, Toldrá F (2004) Histidine dipeptides HPLC-based test for the detection of mammalian origin proteins in feeds for ruminants. Meat Sci 67(2):211–217

Ashbrook SE, Hodgkinson P (2018) Perspective: current advances in solid-state NMR spectroscopy. J Chem Phy 149(4)

Avila Rodríguez MI et al (2018) Collagen: a review on its sources and potential cosmetic applications. J Cosmet Dermatol 17(1):20–26

Awin T, Mediani A, Shaari K, Faudzi SMM, Sukari MAH, Lajis NH, Abas F (2016) Phytochemical profiles and biological activities of Curcuma species subjected to different drying methods and solvent systems: NMR-based metabolomics approach. Ind Crops Prod 94:342–352. https://doi.org/10.1016/j.indcrop.2016.08.020

Bagal-Kestwal DR et al (2019) Properties and applications of gelatin, pectin, and carrageenan gels. In: Bio monomers for green polymeric composite materials, pp 117–140

Ballin NZ (2010) Authentication of meat and meat products. Meat Sci 86(3):577–587

Bartlett S, Davidson W (1992) FINS (forensically informative nucleotide sequencing): a procedure for identifying the animal origin of biological specimens. Biotechniques 12(3):408–411

Battarjee S et al (1999) Preparation of medicinal petroleum jelly using local petroleum waxes. Lubr Sci 12(1):89–104

Bell J, Gillatt P (1994) Standards to ensure the authenticity of edible oils and fats. Aliment Nutr Agric (FAO) 11:29–35

Berrini A et al (2006) Identification of freshwater fish commercially labelled "perch" by isoelectric focusing and two-dimensional electrophoresis. Food Chem 96(1):163–168

Brieskorn C, Sax H (1970) Synthesis of glycyrrhizin acid and glycyrrhetin acid derivatives. Arch Pharm Ber Dtsch Pharm Ges 303(11):905–912

Brodmann PD, Moor D (2003) Sensitive and semi-quantitative TaqMan™ real-time polymerase chain reaction systems for the detection of beef (Bos taurus) and the detection of the family Mammalia in food and feed. Meat Sci 65(1):599–607

Buchgraber M et al (2004) Triacylglycerol profiling by using chromatographic techniques. Eur J Lipid Sci Technol 106(9):621–648

Calvo J et al (2001) Random amplified polymorphic DNA fingerprints for identification of species in poultry pate. Poult Sci 80(4):522–524

Capo RC et al (1998) Strontium isotopes as tracers of ecosystem processes: theory and methods. Geoderma 82(1–3):197–225

Cativiela C et al (2003) Heterogeneous catalysis in the synthesis and reactivity of allantoin. Green Chem 5(2):275–277

Chisholm J et al (2005) The detection of horse and donkey using real-time PCR. Meat Sci 70(4):727–732

Couteau C, Coiffard L (2016) Overview of skin whitening agents: drugs and cosmetic products. Cosmetics 3(3):27

Dalvit C et al (2008) Breed assignment test in four Italian beef cattle breeds. Meat Sci 80(2):389–395

DeNiro MJ, Epstein S (1978) Influence of diet on the distribution of carbon isotopes in animals. Geochim Cosmochim Acta 42(5):495–506

DFG, D. F (2015) "Glycerin." The MAK–collection part. I: MAK value documentations; Wiley-VCH GmbH & Co.: KGaA, Weinheim

Draisci R et al (2000) Quantitation of anabolic hormones and their metabolites in bovine serum and urine by liquid chromatography–tandem mass spectrometry. J Chromatogr A 870(1–2):511–522

Duckett S et al (1993) Effects of time on feed on beef nutrient composition. J Anim Sci 71(8):2079–2088

Dunne P et al (2009) Colour of bovine subcutaneous adipose tissue: a review of contributory factors, associations with carcass and meat quality and its potential utility in authentication of dietary history. Meat Sci 81(1):28–45

Emwas AHM. (2015) The strengths and weaknesses of NMR spectroscopy and Mass spectrometry with particular focus on metabolomics research. In: Bjerrum J (ed) Metabonomics. Methods in Molecular Biology, vol 1277. Humana Press, New York, NY. https://doi.org/10.1007/978-1-4939-2377-9_13

References

Etherington DJ, Sims TJ (1981) Detection and estimation of collagen. J Sci Food Agric 32(6):539–546

Fadzillah NA et al (2017) Authentication of butter from lard adulteration using high-resolution of nuclear magnetic resonance spectroscopy and high-performance liquid chromatography. Int J Food Prop 20(9):2147–2156

Fang G et al (2013) Characterization of oils and fats by 1H NMR and GC/MS fingerprinting: classification, prediction and detection of adulteration. Food Chem 138(2–3):1461–1469

Fatima N et al (2023) Opinions of the imams in the ingredients and manufacturing processes of cosmetics permissible to the Muslim community. J Positive School Psychol 7:801–806

Forrest A, Carnegie P (1994) Identification of gourmet meat using FINS (forensically informative nucleotide sequencing)

Fumière O et al (2006) Effective PCR detection of animal species in highly processed animal byproducts and compound feeds. Anal Bioanal Chem 385:1045–1054

González-Córdova A et al (1998) Detección inmunoquímica de la adulteración de chorizo de cerdo con proteínas de soja: Immunochemical detection of fraudulent adulteration of pork chorizo (sausage) with soy protein. Food Sci Technol Int 4(4):257–262

Hartwig M et al (1997) Physiological quantities of naturally occurring steroid hormones (androgens and progestogens), precursors and metabolites in beef of differing sexual origin. Zeitschrift für Lebensmitteluntersuchung und-Forschung A 205:5–10

Hepburn H, Hepburn H (1986) Composition and synthesis of beeswax. In: Honeybees and wax: an experimental natural history, pp 44–56

Iijima K et al (2006) DNA analysis for identification of food-associated foreign substances. J Food Qual 29(5):531–542

Imanaka H et al (1999) Liposomal linoleic acid is useful as a skin lightening agent. J Soc Cosmet Chem Jpn 33(3):277–282

Iwata H, Shimada K (2012) Formulas, ingredients and production of cosmetics: technology of skin-and hair-care products in Japan. Springer

Iwata H et al (2013) Developing the formulations of cosmetics. In: Formulas, ingredients and production of cosmetics: technology of skin-and hair-care products in Japan, pp 3–19

Jakes W et al (2015) Authentication of beef versus horse meat using 60 MHz 1H NMR spectroscopy. Food Chem 175:1–9

Kawęcka A, Pasternak M (2022) The effect of Slaughter age on meat quality of male kids of the Polish Carpathian native goat breed. Animals (Basel) 12(6):702

Kim J, DellaPenna D (2006) Defining the primary route for lutein synthesis in plants: the role of Arabidopsis carotenoid β-ring hydroxylase CYP97A3. Proc Natl Acad Sci 103(9):3474–3479

Kim D et al (2009) Fatty acid composition and meat quality traits of organically reared Korean native black pigs. Livest Sci 120(1–2):96–102

Kim HK, Choi YH, Verpoorte R (2010) NMR-based metabolomic analysis of plants. Nat Protoc 5:536–549. https://doi.org/10.1038/nprot.2009.237

Kim T et al (2011) Porcine amniotic fluid as possible antiwrinkle cosmetic agent. Korean J Chem Eng 28:1839–1843

Kim YS et al (2018) Effect of DNA extraction methods on the detection of porcine ingredients in Halal cosmetics using real-time PCR. Appl Biol Chem 61:549–555

Kumar N, Pruthi V (2014) Potential applications of ferulic acid from natural sources. Biotechnol Rep 4:86–93

Lee JY et al (2018) Quantitative analysis of lard in animal fat mixture using visible Raman spectroscopy. Food Chem 254:109–114

Lembcke J et al (2005) Rapid quantification of free and esterified phytosterols in human serum using APPI-LC-MS/MS. J Lipid Res 46(1):21–26

Li Z et al (2015) An improved and practical synthesis of tranexamic acid. Org Process Res Dev 19(3):444–448

Li Y et al (2016) Lamellar liquid crystal improves the skin retention of 3-O-ethyl-ascorbic acid and potassium 4-methoxysalicylate in vitro and in vivo for topical preparation. Aaps Pharmscitech 17:767–777

Lin Y, Yan Y (2012) Biosynthesis of caffeic acid in Escherichia coli using its endogenous hydroxylase complex. Microb Cell Fact 11:1–9

Linster CL, Van Schaftingen E (2007) Vitamin C: biosynthesis, recycling and degradation in mammals. FEBS J 274(1):1–22

Lockley A, Bardsley R (2000) DNA-based methods for food authentication. Trends Food Sci Technol 11(2):67–77

Lopes BdM et al (2015) Xanthan gum: properties, production conditions, quality and economic perspective. J Food Nutr Res 54(3):185–194

López-Andreo M et al (2006) Evaluation of post-polymerase chain reaction melting temperature analysis for meat species identification in mixed DNA samples. J Agric Food Chem 54(21):7973–7978

Lukitaningsih E et al (2012) Quantitative analysis of lard in cosmetic lotion formulation using FTIR spectroscopy and partial least square calibration. J Am Oil Chem Soc 89:1537–1543

Masoum S et al (2007) Application of support vector machines to 1 H NMR data of fish oils: methodology for the confirmation of wild and farmed salmon and their origins. Anal Bioanal Chem 387:1499–1510

Mathew VN (2014) Acceptance on Halal food among non-Muslim consumers. Procedia Soc Behav Sci 121:262–271

Mortas M et al (2022) Adulteration detection technologies used for Halal/kosher food products: an overview. Discov Food 2(1):15

Nagai J, Bloch K (1965) Synthesis of oleic acid by Euglena gracilis. J Biol Chem 240(9):3702–3703

Nair V et al (2006) Determination of stigmasterol, β-sitosterol and stigmastanol in oral dosage forms using high performance liquid chromatography with evaporative light scattering detection. J Pharm Biomed Anal 41(3):731–737

Netscher T (2007) Synthesis of vitamin E. Vitam Horm 76:155–202

Petrakis EA, Cagliani LR, Polissiou MG, Consonni R (2015) Evaluation of saffron (Crocus sativus L.) adulteration with plant adulterants by1H NMR metabolite fingerprinting. Food Chem 173:890–896. https://doi.org/10.1016/j.foodchem.2014.10.107

Prache S et al (2003) Persistence of carotenoid pigments in the blood of concentrate-finished grazing sheep: its significance for the traceability of grass-feeding. J Anim Sci 81(2):360–367

Precht D (1992) Detection of foreign fat in milk fat. I. Qualitative detection by triacylglycerol formulae. Z Lebensm Unters Forch 194:1

Qin X, Zhong J (2016) A review of extraction techniques for avocado oil. J Oleo Sci 65(11):881–888

Rahman MA et al (2003) Synthesis of palm kernel oil alkanolamide using lipase. J Oleo Sci 52(2):65–72

Sasazaki, S., et al. (2004). "Development of breed identification markers derived from AFLP in beef cattle." Meat Sci 67(\): 275–280

Schagen SK (2017) Topical peptide treatments with effective anti-aging results. Cosmetics 4(2):16

Schlossman ML, Mccarthy JP (1978) Lanolin and its derivatives. J Am Oil Chem Soc 55(4):447–450

Schmidt O et al (2005) Inferring the origin and dietary history of beef from C, N and S stable isotope ratio analysis. Food Chem 91(3):545–549

Seretis A, Tsiakaras P (2016) Hydrogenolysis of glycerol to propylene glycol by in situ produced hydrogen from aqueous phase reforming of glycerol over SiO2–Al2O3 supported nickel catalyst. Fuel Process Technol 142:135–146

Shui T et al (2017) Synthesis of sodium carboxymethyl cellulose using bleached crude cellulose fractionated from cornstalk. Biomass Bioenergy 105:51–58

Simontacchi C et al (1999) Accuracy in naturally occurring anabolic steroid assays in cattle and first approach to quality control in Italy. Analyst 124(3):307–312

Sugibayashi K et al (2019) Halal cosmetics: a review on ingredients, production, and testing methods. Cosmetics 6(3):37

References

Sze JH et al (2016) Biotechnological production of hyaluronic acid: a mini review. 3 Biotech 6:1–9

Szűcs S et al (2006) Method validation for the simultaneous determination of fecal sterols in surface waters by gas chromatography-mass spectrometry. J Chromatogr Sci 44(2):70–76

Tang F et al (2019) Magnetic resonance applications in food analysis. Annu Rep NMR Spectrosc 98:239–306

Verkaar E et al (2002) Differentiation of cattle species in beef by PCR-RFLP of mitochondrial and satellite DNA. Meat Sci 60(4):365–369

Windarsih A, Rohman A, Swasono RT (2019) Application of 1H-NMR based metabolite fingerprinting and chemometrics for authentication of Curcuma longa adulterated with C. heyneana. J Appl Res Med Aromat Plants 13:100203. https://doi.org/10.1016/j.jarmap.2019.100203

Xiong P et al (2015) Design, synthesis, and evaluation of genistein analogues as anti-cancer agents. Anti-Cancer Agents Med Chem (formerly current medicinal Chemistry-anti-cancer agents) 15(9):1197–1203

Yamano Y, Ito M (2007) Total synthesis of capsanthin and capsorubin using Lewis acid-promoted regio-and stereoselective rearrangement of tetrasubsutituted epoxides. Org Biomol Chem 5(19):3207–3212

Yu X et al (2017) Simple synthesis hydrogenated castor oil fatty amide wax and its coating characterization. J Oleo Sci 66(7):659–665

Yusuf E, Yajid MSA (2017) Related topic: Halal cosmetics. In: Skin permeation and disposition of therapeutic and cosmeceutical compounds, pp 101–107

Zeleny R, Schimmel H (2002) Sexing of beef—a survey of possible methods. Meat Sci 60(1):69–75

Zhang P et al (2014) Bioactivity and chemical synthesis of caffeic acid phenethyl ester and its derivatives. Molecules 19(10):16458–16476

Zhen Z et al (2015) Surface modification by natural biopolymer coatings on magnesium alloys for biomedical applications. In: Surface modification of magnesium and its alloys for biomedical applications. Elsevier, pp 301–333

Zia KM et al (2017) A review on synthesis, properties and applications of natural polymer based carrageenan blends and composites. Int J Biol Macromol 96:282–301

Conclusion

Food authentication is a significant concern in global food production. The primary goal of food authentication is to ensure food safety, quality, and origin, thereby meeting legislative requirements adequately and reliably.

The Muslim population accounts for approximately 25% of the world's population, necessitating adherence to Islamic dietary laws for food ingredients and processing prior to consumption. It is very challenging for Muslim consumers to avoid food or ingredients that are forbidden. Therefore, the development of rapid and cost-effective Halal authentication is crucial. Halal authentication is the process of verifying a food product's compliance with Islamic law.

Chemometric techniques serve as effective tools for classifying and predicting various types of food. The outline of Halal food authentication using FTIR combined with chemometrics is presented. This book aims to broaden the concept of Halal food and elaborate on the use of FTIR in conjunction with chemometrics for Halal authentication.

Features
- Muslim consumers often face challenges in avoiding foods or ingredients that are not permissible under Islamic law.
- Reliable, rapid, and cost-effective Halal authentication methods are therefore essential.
- Chemometric tools provide effective classification and prediction of food authenticity.
- This book discusses the use of FTIR combined with chemometrics as a framework for Halal food authentication.

The book will not only provide the latest insights into the connection between Halal analysis and authentication but also offer concrete and comprehensive information on FTIR and chemometrics for Halal authentication.

Index

A
Additives, 13–15, 17, 48, 61, 62, 121, 126, 134, 135, 137
Advanced analytical techniques, 20, 175
Applications, 16, 21, 24, 25, 27, 30–32, 34, 35, 40, 48, 62, 67, 69, 72, 73, 79–82, 89, 90, 97, 103, 104, 109, 115, 116, 122, 128, 132, 134, 138, 150, 151, 171
Authentication, 3, 19–36, 41, 55, 165–175
Authenticity, 21, 22, 40, 41, 55, 56, 58, 64, 78, 89, 106, 108, 118, 121, 127, 129, 134, 135, 138, 141, 145, 147, 150, 151, 166, 169, 171
Awareness, 9, 11, 12, 165, 167

C
Chemometrics, 22–24, 39–41, 43, 44, 46, 58, 76–78, 80–85, 87, 89, 90, 93–97, 99–101, 103–111, 115, 116, 120–123, 125–129, 131–135, 138, 141, 142, 145–147, 150
Consumers, 3, 9, 10, 12, 13, 15, 16, 19, 41, 55, 56, 58, 59, 61, 78, 87, 108, 127, 129, 134, 135, 138, 150, 151, 165, 167
Cosmetics ingredient authentication, 172–173

F
Food authentication, 1, 19, 24, 25, 28–30, 39, 41–46, 48, 49, 55–151, 171
Food fraud, 40, 59, 62, 146, 150

Food safety, 6, 12, 13, 19, 55, 78, 151
FTIR spectroscopy, 42–44, 48, 55–151, 175

H
Halal, 1, 19–36, 42, 55, 165
Halal authentication of food, 173
Halal foods, 1–17, 19, 20, 30, 39–49, 55–151, 170, 171
Hygiene, 5, 9, 10

L
Logistics, 13

M
Machine learning (ML), 46, 48, 49, 83, 116
Meat adulteration, 87, 89, 95, 103, 146
Meat authentication, 42–44, 48, 85, 146, 165–169
Meat geography, 165–166
Meat origin, 166–167

P
PLS-DA, 80, 81, 83, 84, 104, 107, 109, 115, 116, 118, 120, 122, 132–134, 149
Principal component analysis (PCA), 44, 47, 77–82, 84–87, 89, 90, 93–111, 115–118, 120–123, 126, 128, 132–135, 138, 139, 141–143, 145–149

R
Religious slaughter, 5–7

S
Slaughter, 1, 3, 5–9, 165, 167

Species authentication, 24, 28
Supply chains, 13, 14, 62, 63, 150, 151

U
Univariate and multivariate, 24, 44

MIX
Papier aus verantwortungsvollen Quellen
Paper from responsible sources
FSC® C105338

If you have any concerns about our products,
you can contact us on
ProductSafety@springernature.com

In case Publisher is established outside the EU,
the EU authorized representative is:
**Springer Nature Customer Service Center GmbH
Europaplatz 3, 69115 Heidelberg, Germany**

Printed by Libri Plureos GmbH
in Hamburg, Germany